the sun
is a star

the sun is a star

a voyage through the universe

Dick Frizzell
with Samantha Lord

MASSEY UNIVERSITY PRESS

For Coco Frizzell,
who gave me the idea.

And for my artist friends who helped. I can't imagine how I thought I was going to illustrate this epic on my own. Fortunately I came up with the wizard wheeze of conning all my highly creative friends and family into doing the job for me. It was like herding cats, but we did it, and I don't think I've been involved in a project with such rich and rewarding results. Hats off all round, I think.

contents

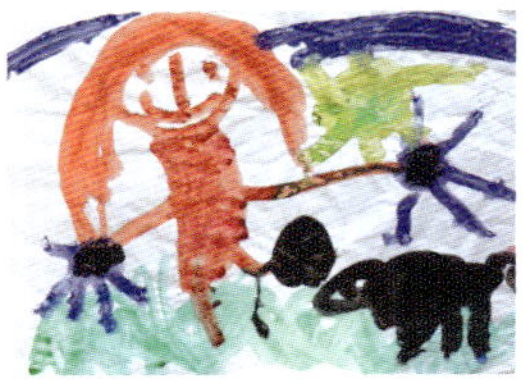

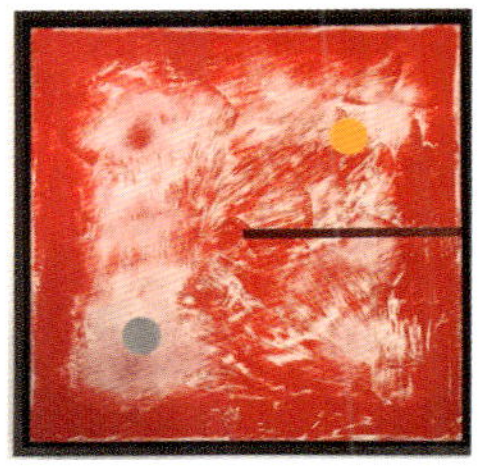

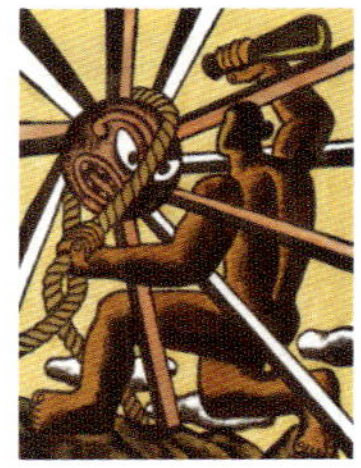

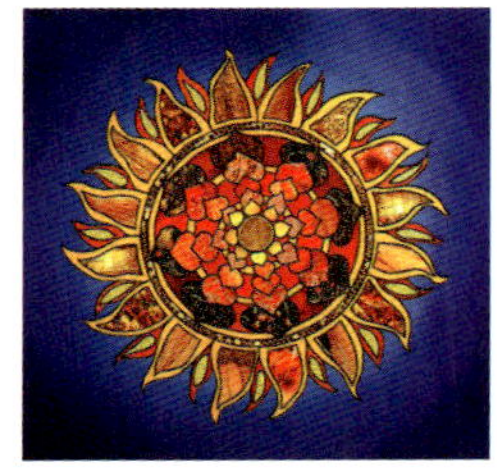

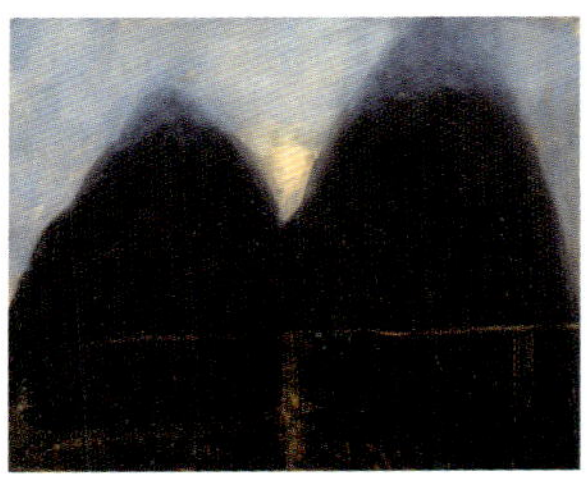

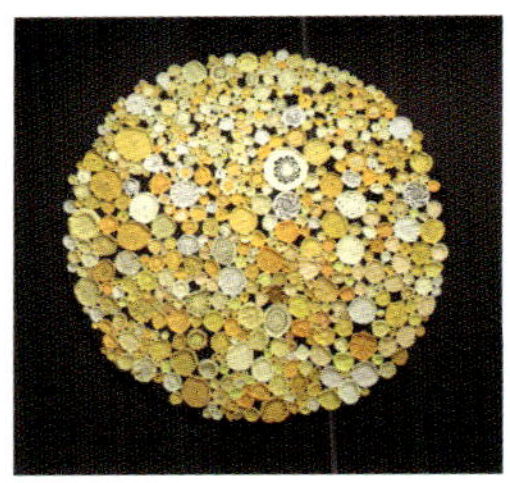

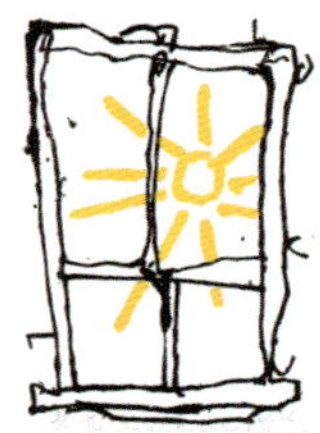

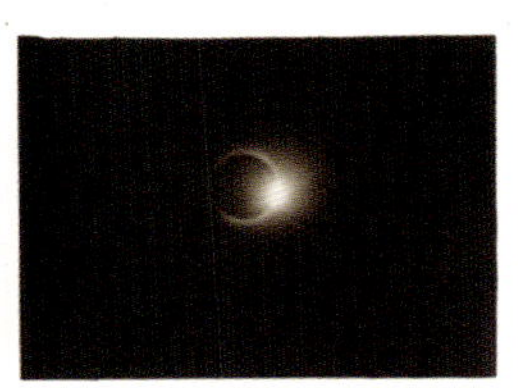

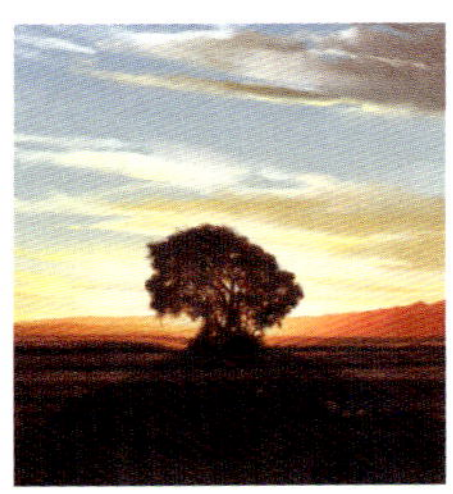

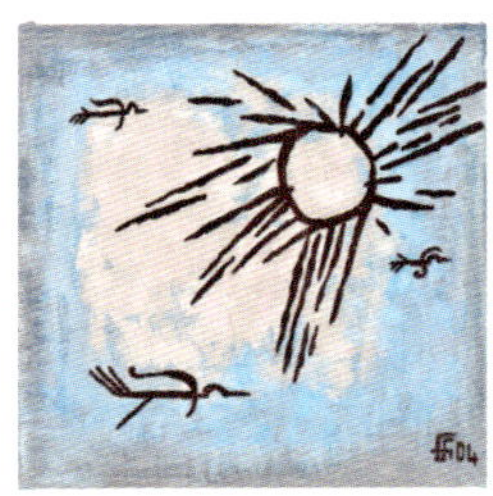

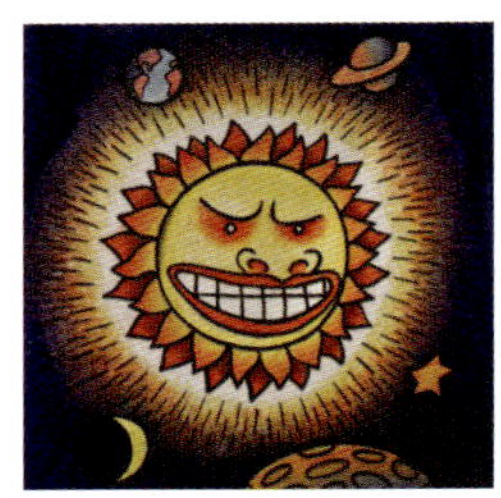

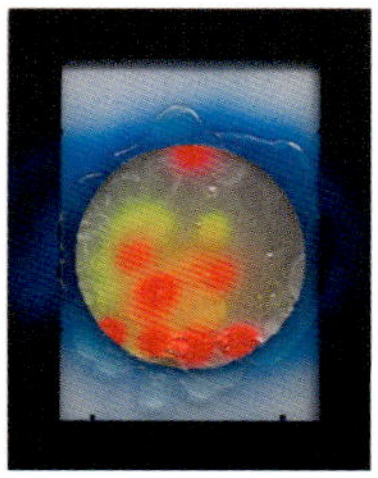

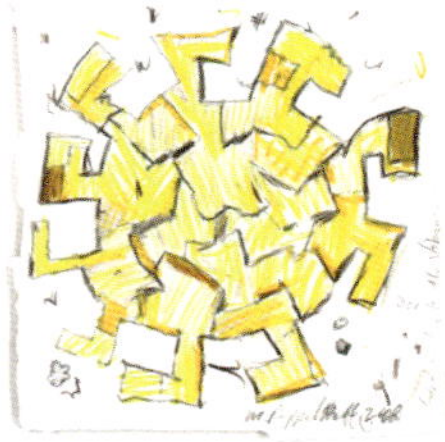

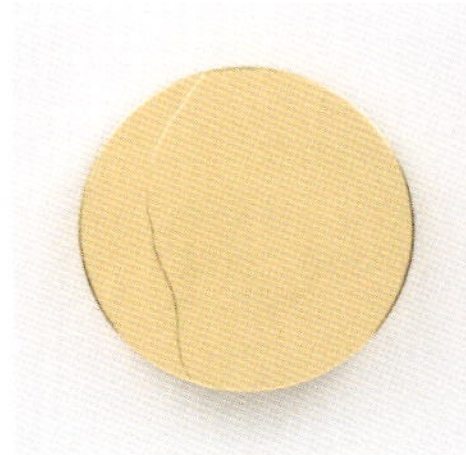

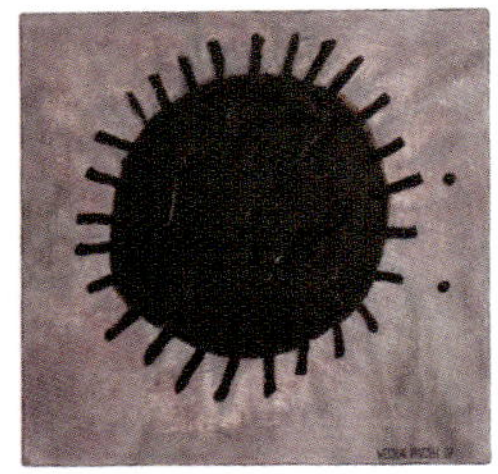

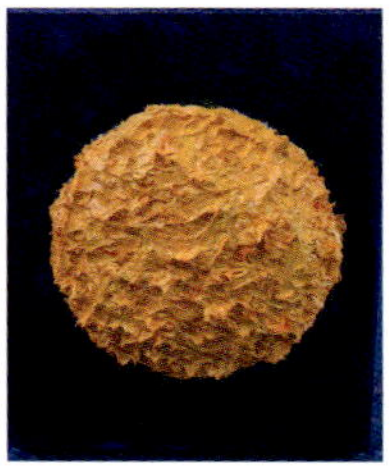

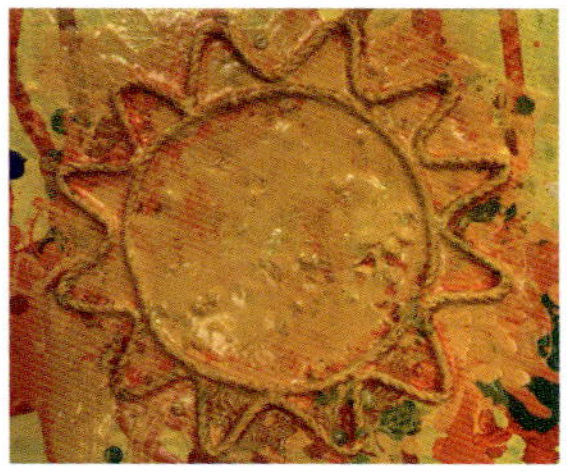

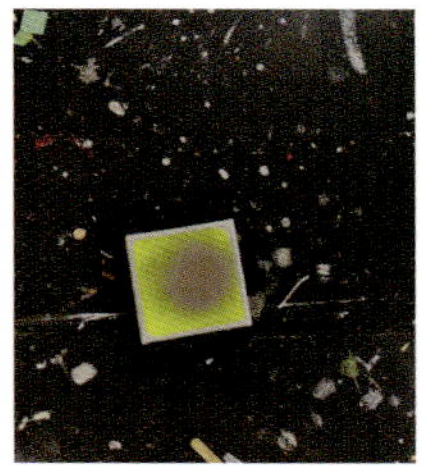

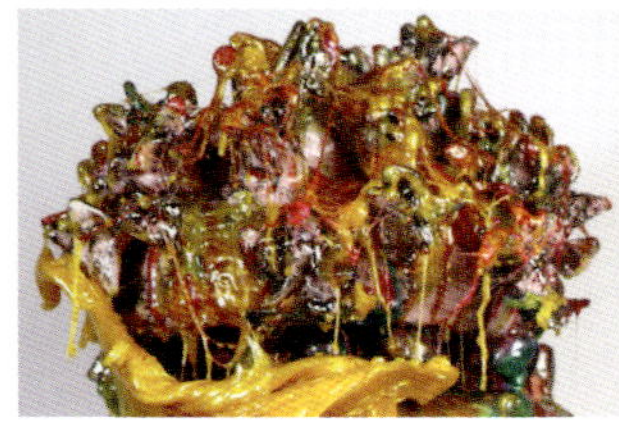

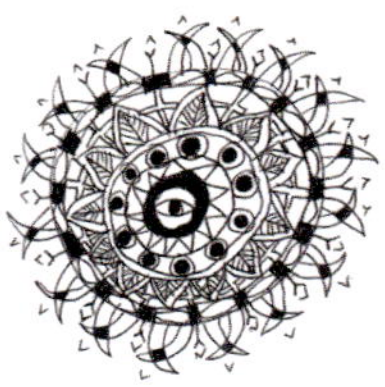

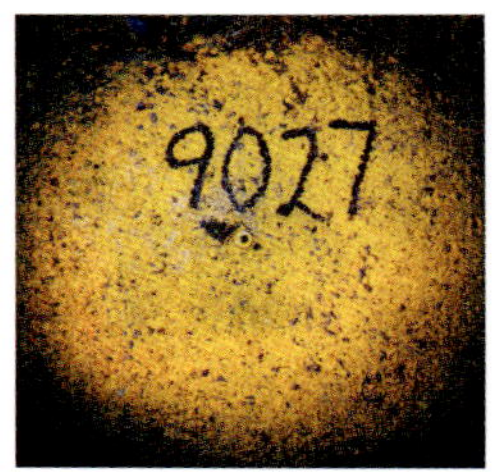
9027

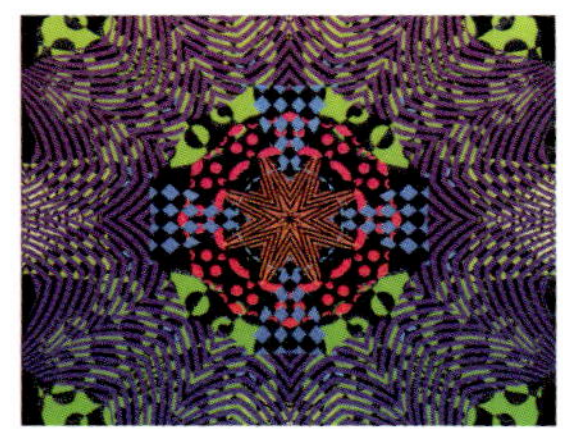

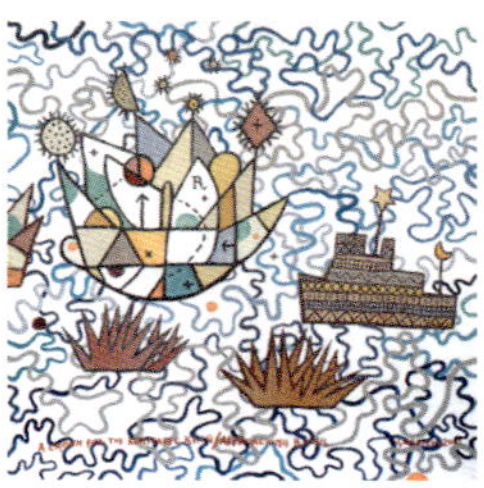

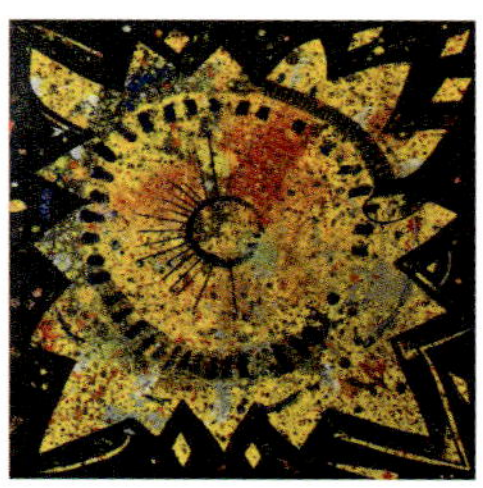

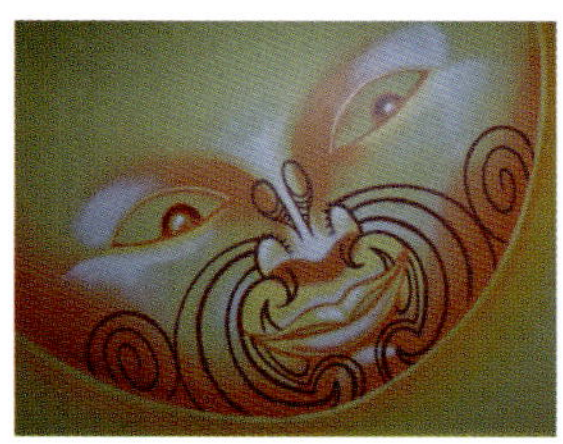

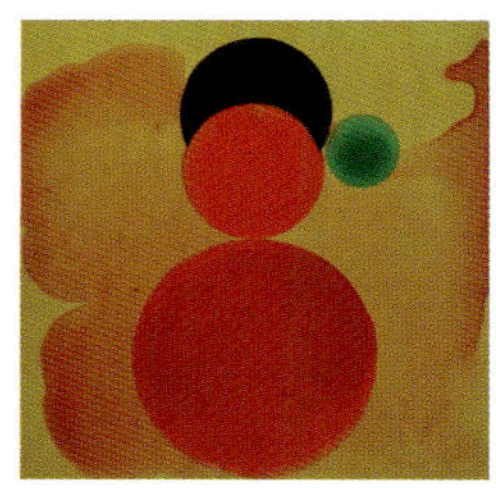

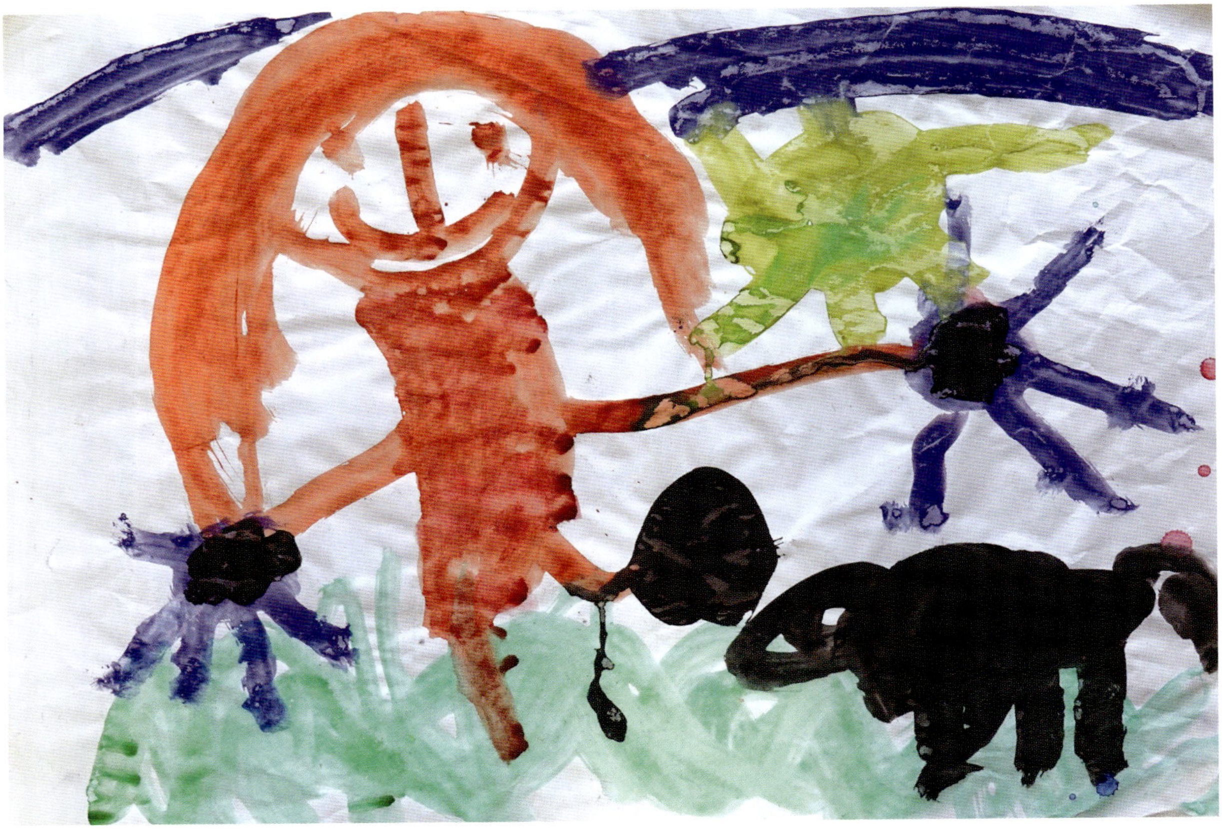

Coco Frizzell, *Me, Modee and the Sun*, 2000

how this book began

Somewhere in my last book I told the story about my granddaughter, Coco Frizzell, seven years old, running into my studio one day after school on her little pea-stick legs and breathlessly announcing 'Dicky, Dicky, did you know that the sun is a star?'

She couldn't wait to pass this astonishing revelation on to me because she knew that I was interested in this sort of stuff. And she was right. I am, and I have been since the early 1960s, when the Big Bang theory became the most popular explanation for the origin of the universe. I was hooked. I was on to it. And I've stayed on to it ever since.

Since then I've had 60 years of constant consumption of popular scientific literature — you'd think I would have learnt something after all that time. Well, I certainly knew that the sun is a star, and when Coco made that highly excitable proclamation I was surprised and delighted to think that cosmology had become so much a part of everyday conversation that it was being taught to seven-year-olds.

Obviously it wasn't a big leap for me to imagine Coco's simple but profound idea expanding into a book. And here it is: *The Sun Is a Star* — everything I know about the universe and everything I've learnt about it since I started writing this book. It started out as a book for a seven-year-old, but I think I ended up writing it for a 77-year-old.

what?
no diagrams?!

As you read, you will notice that I've used no diagrams in this book. I've decided, after a lifetime of looking at them, that science diagrams set off instant alarm bells in the mind of the average reader (like, 'Uh-oh, I'm about to be hornswoggled'). They tend to create more confusion than clarity.

A classic example — to me anyway — is the standard attempt to illustrate the warp and weft of space–time. This is usually a stretched-out net thing that tapers off towards the top of the page like the opening plot summary in *Star Wars*. As you sit there looking over this geometric arrangement it's hard not to feel as if you're looking down at the carpet: flat, flat, flat.

But what if I were to shrink you down like Ant Man until you were *in* the weave of that carpet? You would then be unaware of the flatness. You would also be aware that your presence in the weave has disturbed its density — in all directions. →

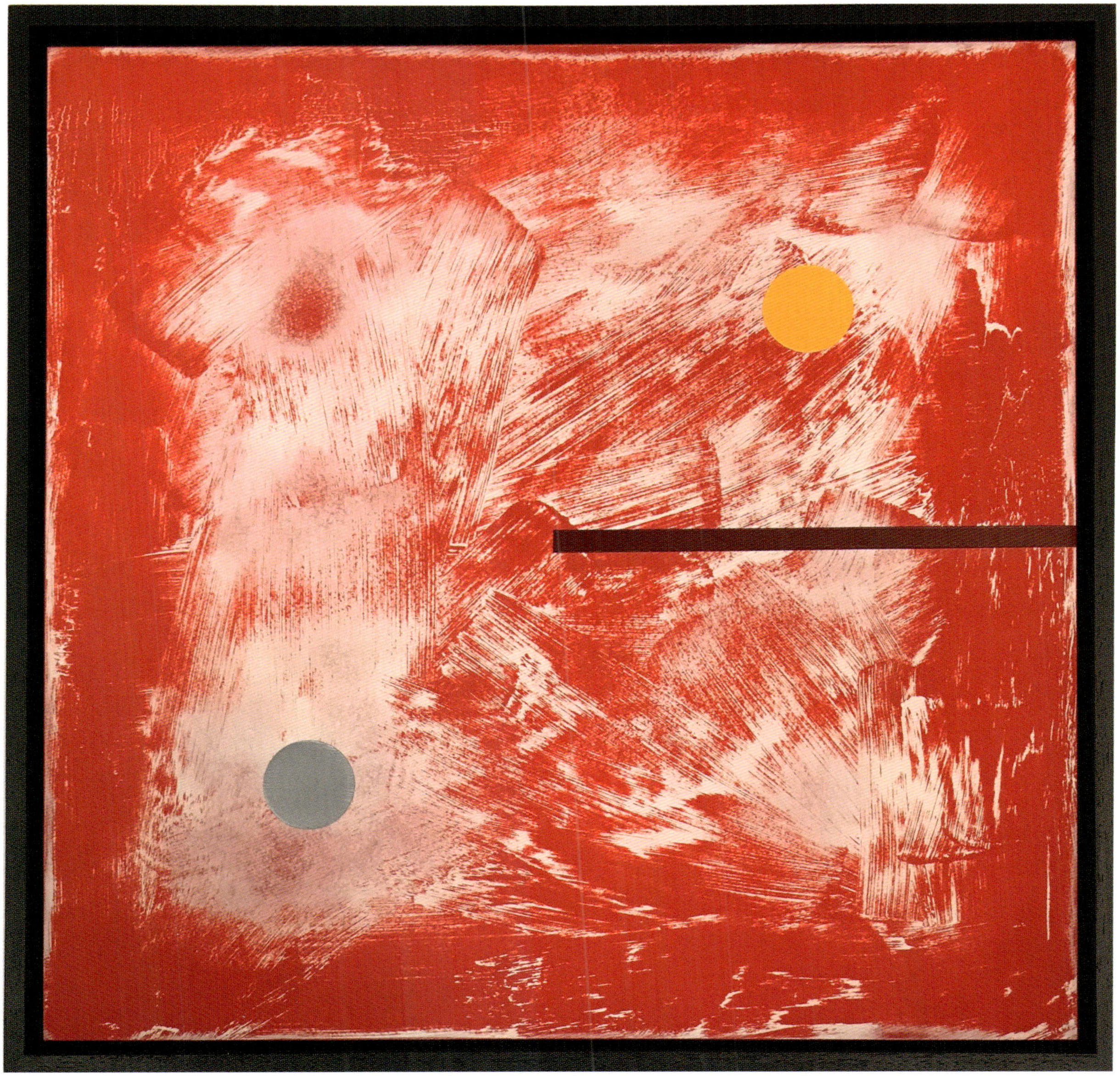

SUN
RISE
UNDER
Wayne Youle, *GROUND*, 2020, oil and acrylic on board,
373 × 373 mm, courtesy of the artist and Suite Gallery

So, you are now inside the diagram, and not on the next page looking at a picture of a bowling ball making a dent in a trampoline, which is another favourite of diagram makers. This visual trick is designed to point out that gravity doesn't really act like a magnet. The 'attraction' of gravity is a result of a mass displacing the fabric of space in such a way that any object near this 'dent' falls towards the thing (the mass) doing the denting.

The thing that the bowling ball diagram has trouble describing is the encompassing nature of the dent. And that's my point. It's more like shoving an orange into a Christmas stocking than placing a bowling ball on a trampoline. And neither diagram is good for explaining how objects in space interact with each other. The only thing that stops everything in space rolling together into one massive clump is the centrifugal force of each individual object's orbit.

And then there's the dreaded 'cone' diagram. I'm never sure just what these cones are trying to represent . . . One of the more persistent impressions is that of pressing a glass up to the wall to try to hear what the people in the neighbouring unit are up to. You know they're up to *something* but you can't figure out what it is.

So, no diagrams. But hopefully some sense.

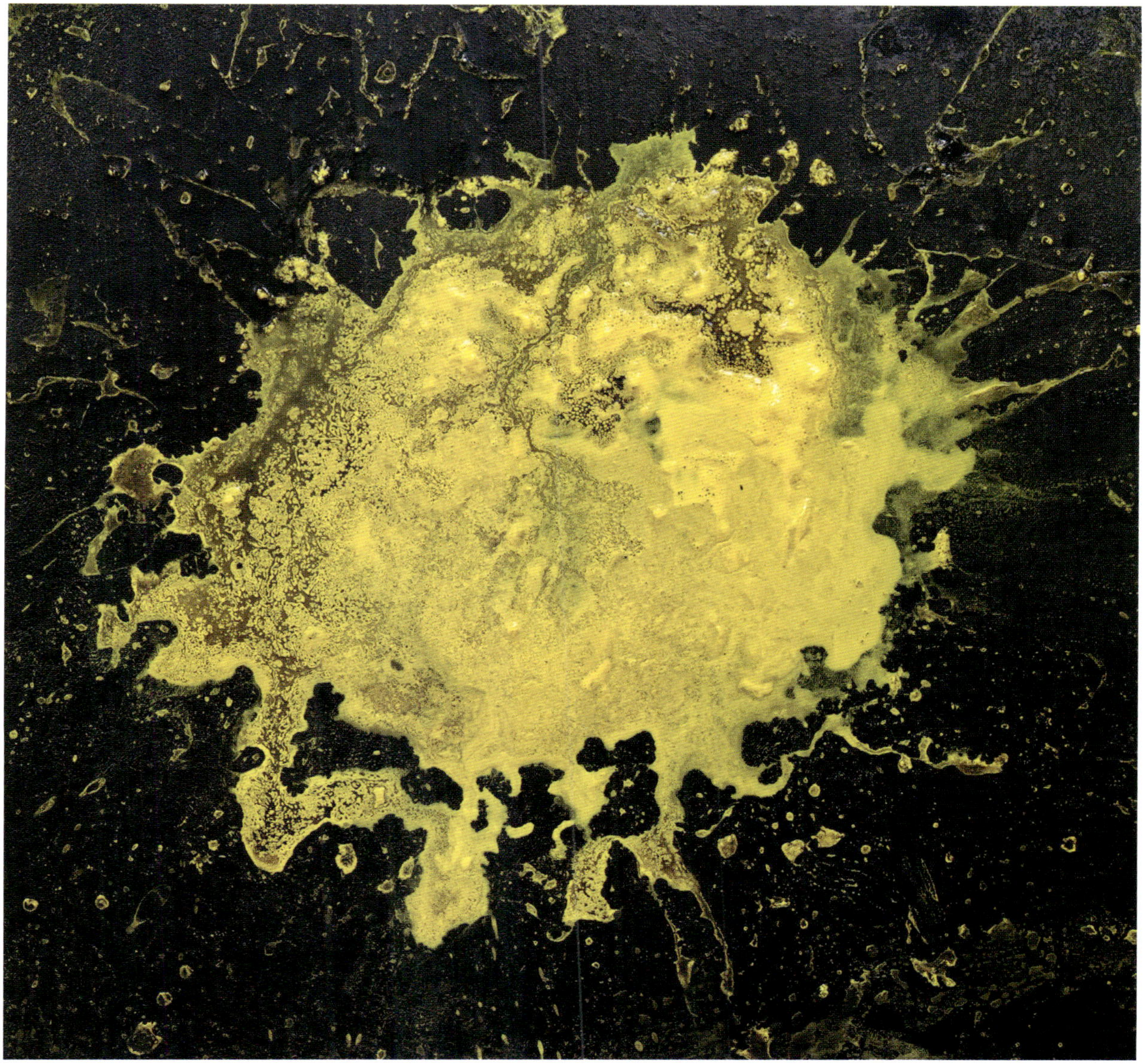

Dick Frizzell, *Kablooey*, 2021, acrylic and enamel on cardboard

the sun is a star

This is a simple fact that a lot of people seem to find hard to grasp. How could it possibly be a star? It's hot, it's big, it's yellow and it sits up there in a wide blue sky. Stars — surely? — are cold, small, white and hover about in the blackness of the night sky.

Not so! If you were able to rocket away to one of those stars, you would find that it would look like the sun and 'our' sun would look like a star!

As far as we know (which isn't very far, to be honest), the sun is the only star in the galaxy that's orbited by a planet full of people calling it names. And here are some of those names, many of which start with S: Soleil, Sol, Sonne, Zon, Sole and Qorraxdu. That last one is from the Somali language. Our own Māori name, Rā, is pretty good. Strong and concise.

And while we're naming things, here are the names of some of the other stars we can see from Earth: Sirius, Canopus, Vega, Procyon and Betelgeuse. These names were made up by ancient Greek or Arabic stargazers looking up at the same stars we look up at today. If there is anyone on the other side of the galaxy looking up at these same stars, they will have completely different names for them. They will also have another name for our star — if they've even noticed it yet, because they do have between 100 billion and 400 billion stars to sort through. Seriously. That's how many stars there are out there, just in our Milky Way Galaxy.

Here's another interesting fact: the light from these stars has taken so long to reach us that some of them are now dead and burnt out, but we don't know it yet.

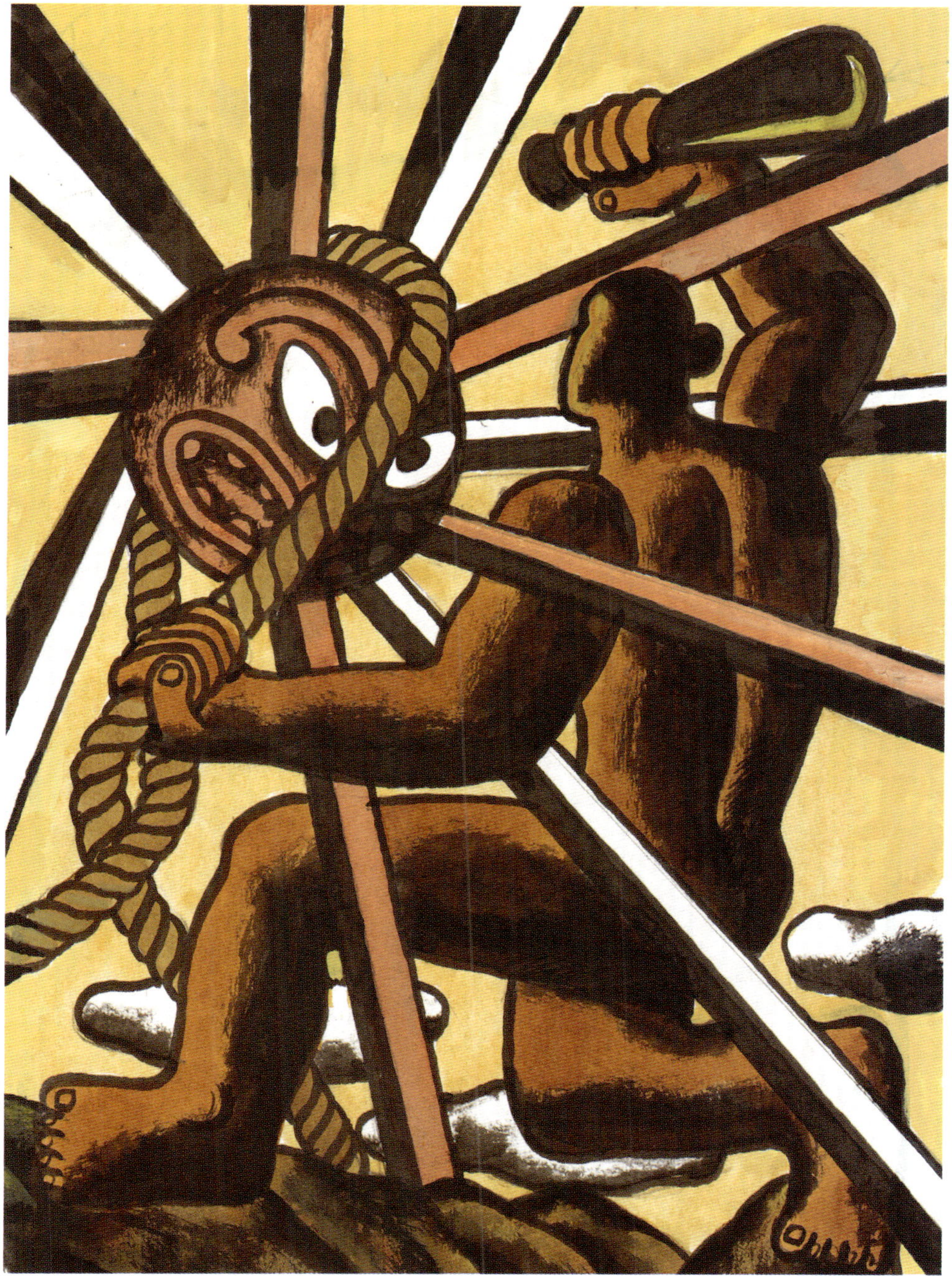

Dick Frizzell, *Maui Captures the Sun*, 2000, study for mural at Te Manawa Art Gallery, Palmerston North

hotter than toast

Our star is not very special in the grand scheme of the universe, but it is still an incredible thing.

For a start, it's hot: very, very hot — in fact, the surface of the sun is an impressive 5600 degrees Celsius. Although lightning is five times hotter! And a supernova — an exploding star — is 300 times hotter than that, at one billion degrees Celsius.

How could anything be that hot? The supernova obviously wins the heat competition, but if you could drill down to the heart of the sun you would find that it is still pretty damn hot — so hot that I'm starting to regret getting into this comparison thing. The temperature in the sun's core is 15 million degrees Celsius. Your cup of tea is about 80 degrees Celsius. So now you know.

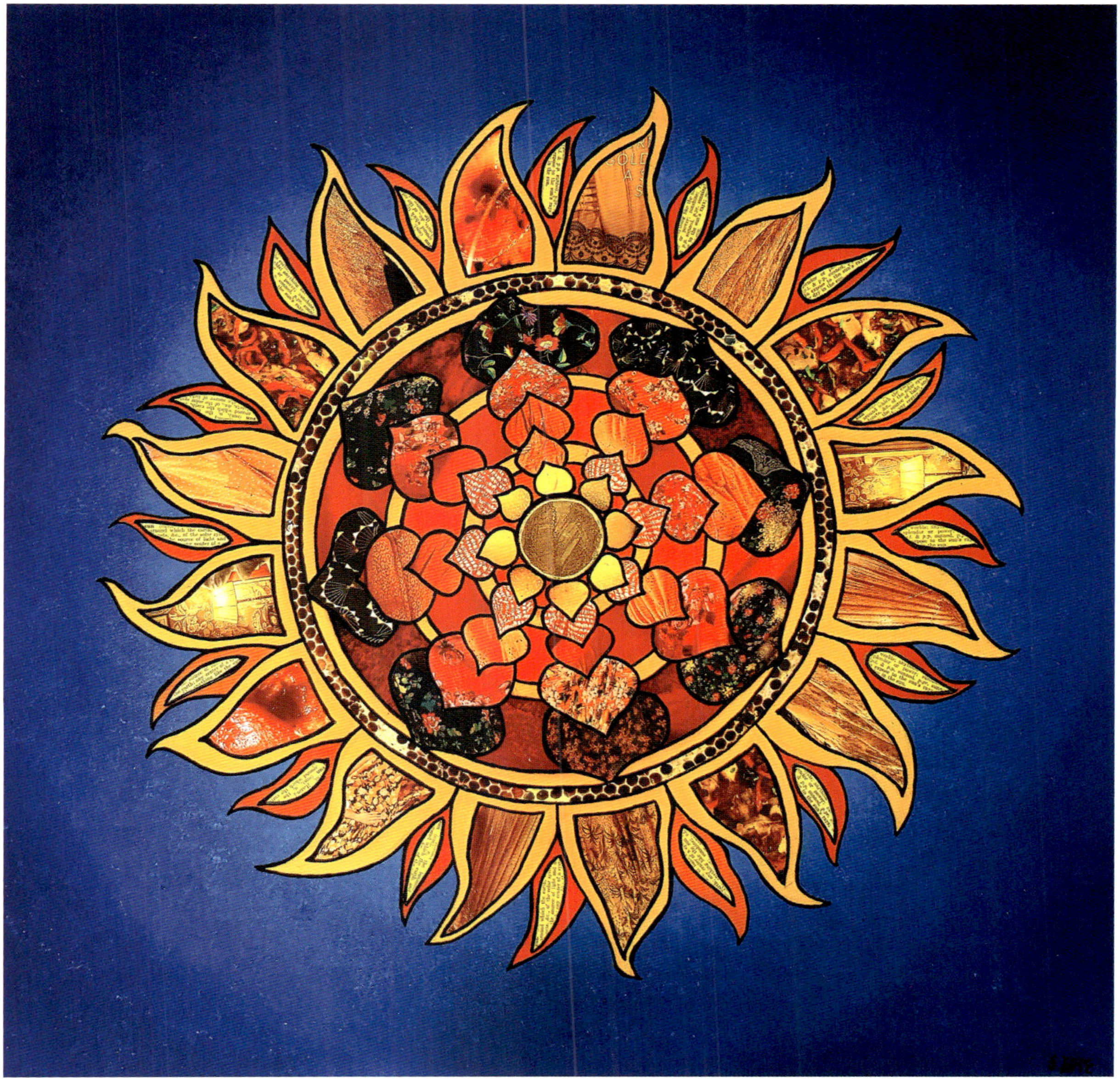

Susan Edge, *Mandala Sun*, 2021, acrylic and collage on canvas, 760 × 760 mm

bigger than big

What else? Well, the sun is huge: hugely huge. It's bigger than all the planets in our solar system put together, and some of those are mind-bogglingly big. Jupiter, for example, is bigger than all the mass of the remaining planets — and the sun could swallow Jupiter without a burp. Comparing the planet Earth with the sun is like comparing a grain of sand with a beach ball — a *big* beach ball.

If you could fly a plane around the sun at normal plane altitude, without getting burnt to a crisp, it would take you about eight months, whereas it takes only about two and a half days to fly around the Earth. And it takes twelve months for the Earth to get around the sun — a whole year. So, given that the Earth's orbit is so much further out than our imaginary plane's normal altitude, how come we only take four months longer to get around the sun?

It's the speed. The Earth is zooming around the sun at 30 kilometres a second, or 110,000 kilometres an hour, while that Air New Zealand plane is only doing about 800 kilometres an hour. No contest really!

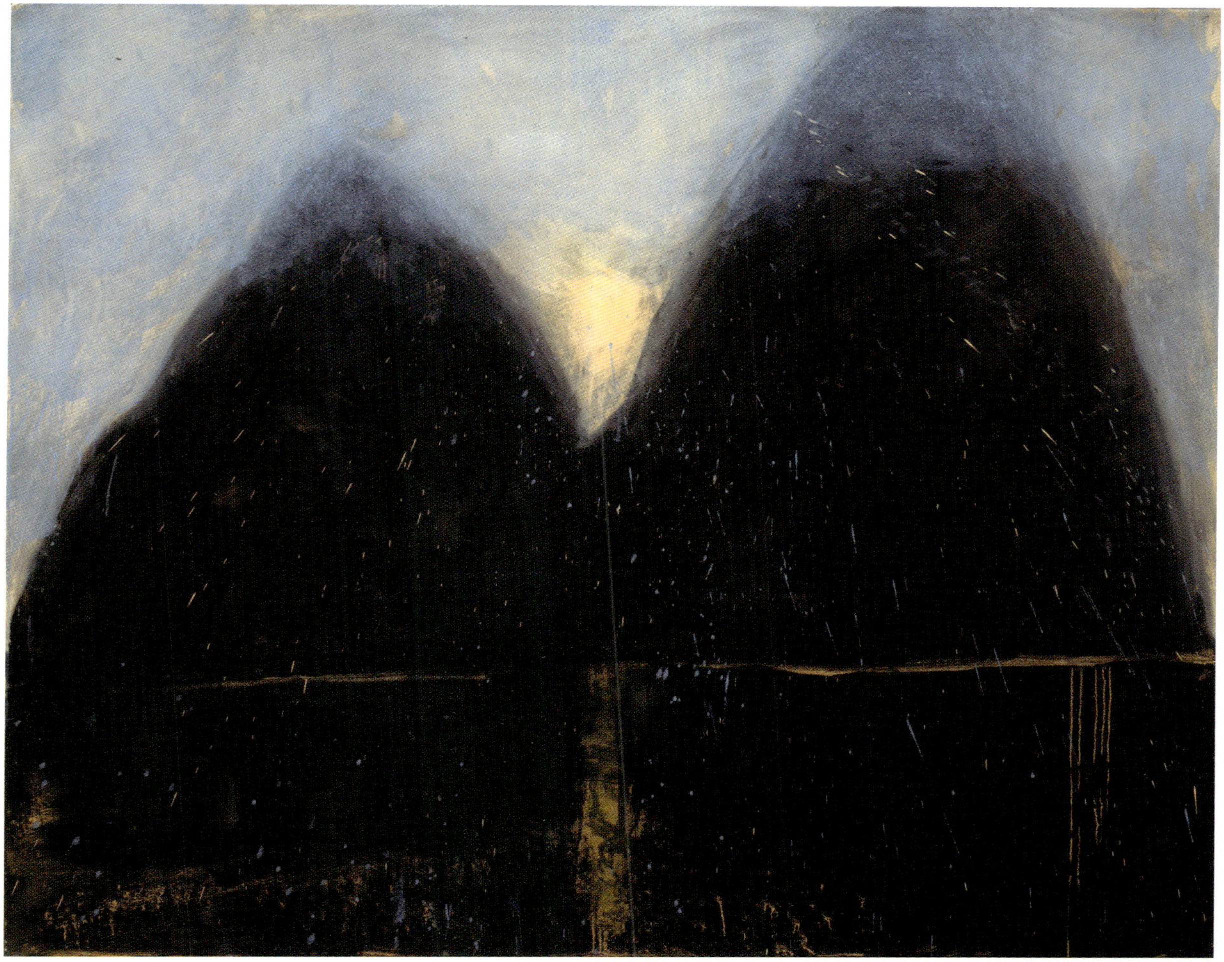

Peter Kingston, *Sacred House*, 2008, oil and gouache on canvas, 1190 × 1340 mm

moonlight is sunlight

We know more or less how moonlight works: the light of the moon is actually the sun's light reflecting off the moon's face. But how does the moon come and go the way it does?

A curious arrangement of vectors and orbits means that the moon always presents the same face to the Earth at all times (and it's curious that it does look like a face, isn't it?). No matter what the time of day or night, the man in the moon never takes his eyes off the Earth.

But once a month he seems to totally disappear, bit by bit. How does this happen? Where does he go?

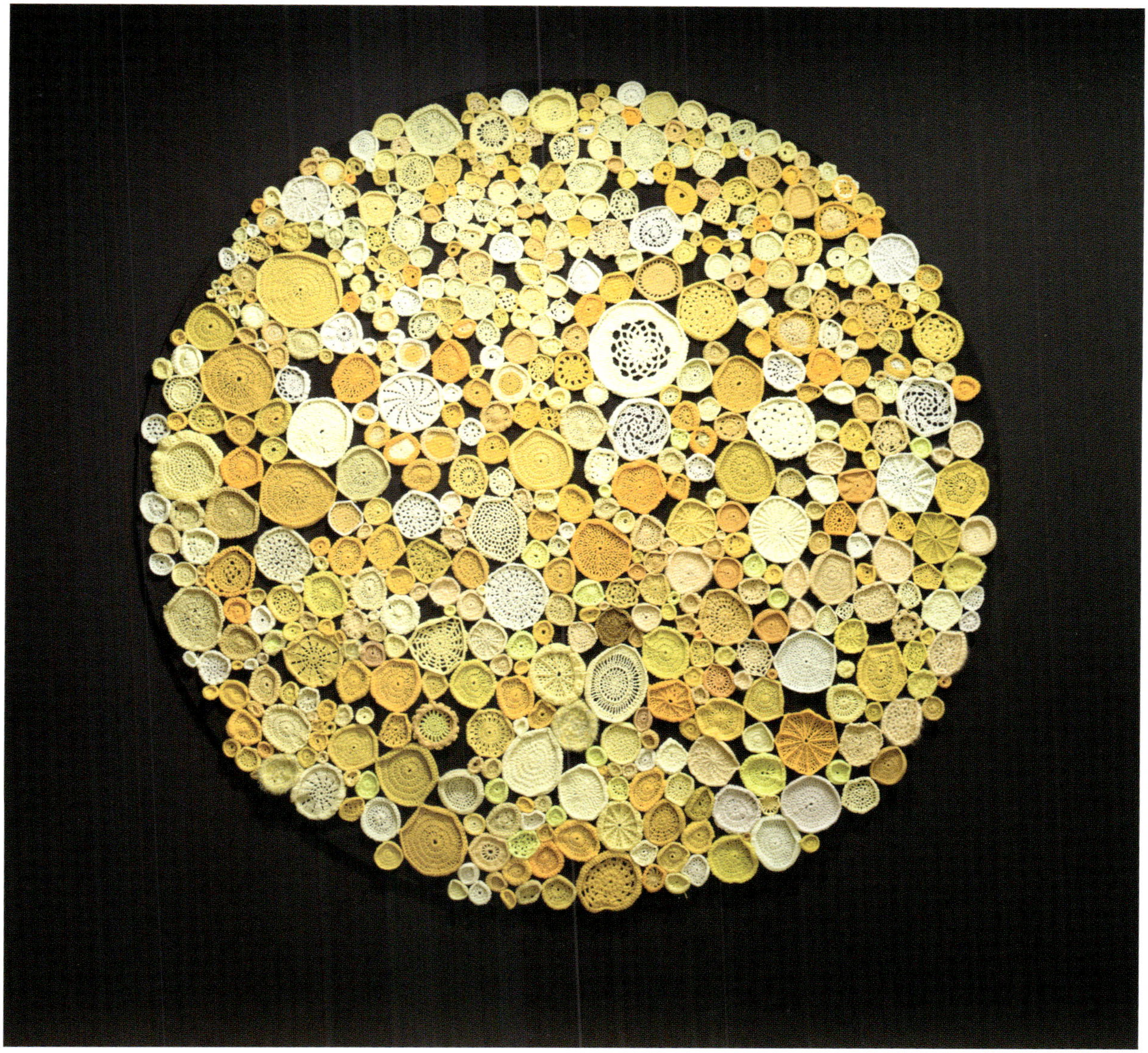

Ani O'Neill, *Crater — Creator*, 2017, collection of Christchurch Art Gallery Te Puna o Waiwhetū, gift of the artist, 2019

no dark side

Here's a simple little parlour game that explains quite clearly how the phases of the moon happen and where that man goes. Ideally, the darker the parlour, the better this works.

Ask a friend to stand in the middle of the room, to represent the Earth. Now turn on a wall light or standard lamp on one side of the room — this represents the sun. You are the moon. Position yourself on the opposite side of the room, facing your friend and in a spot where she isn't getting in the way of the light (I'll explain in a minute how the Earth manages to do that).

Ask your friend (the Earth) what she sees, and of course she'll tell you that she sees your face totally lit by the lamp. Now walk in an arc (a circle) to the left around your friend, all the while keeping your face towards her. Stop a quarter of the way round and ask again what she can see. Of course, she'll tell you that half your face is now lit up. This is the moon's first quarter.

Now — and this is the interesting bit — keep sidling in the circle until the back of your head is towards the lamp, and ask your friend what she sees now. If my simple game has worked, she will see your face in complete shadow. (And this is why, despite the name of Pink Floyd's famous album, there is no dark side of the moon.) →

Dick Frizzell and Martin Poppelwell, *The Sun is in the Window*, 2021, silkscreen print

At this point in the game (and when 'seen' against the night sky), you as the moon have 'disappeared'.

The minute the moon does this disappearing act we call it the New Moon, which seems a bit contradictory but that's the way it works. If you keep circling around your patient friend, the light will start to catch your face again until, eventually — in 15 days — you will arrive back where you started with full lamplight in your face.

You've just acted out all eight phases of the moon: full moon, not-so-full (called 'gibbous' — isn't that a great word?), first quarter, first crescent (going), new (gone), second crescent (coming), second quarter, gibbous again, and back to full.

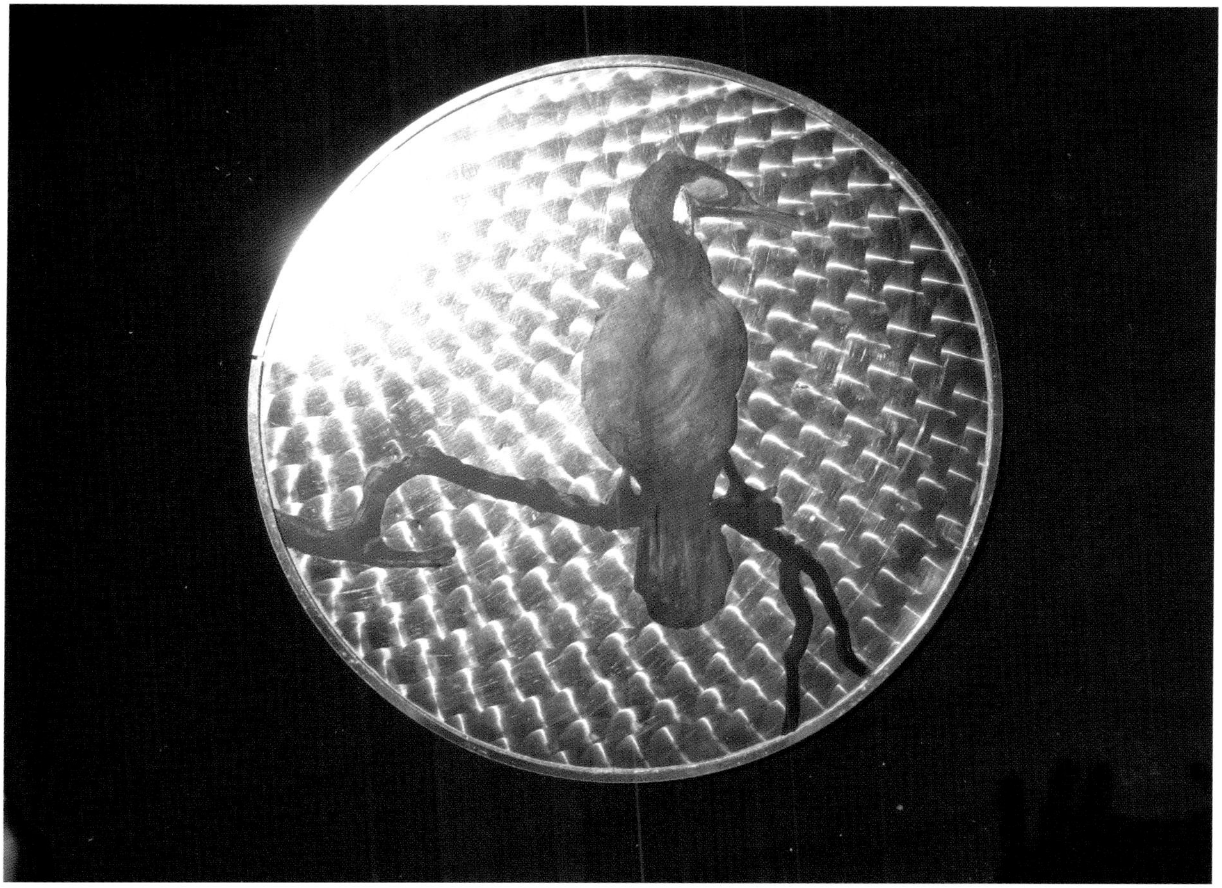

Simon Shepheard, *Sun dreaming fish scales*, 2017, primer and acrylic paint
on polished steel café table top, 650 mm diameter

it’s a revolution

In the month that the moon takes to orbit the Earth, it also makes a complete revolution in itself. How could this be?, you might ask — we never see it turning round!

Here’s how you can figure it out: sidle round in a circle facing the centre of a room, like the moon going round the Earth. Then sidle again standing on the spot in the centre of the room and staring down at your feet. You’ll be surprised to find that you’ve turned a full 360 degrees!

Am I overthinking this? Possibly, but because the man in the moon is so determined to keep an eye on us as he circles about us, it’s counter-intuitive to think of him revolving as he does it!

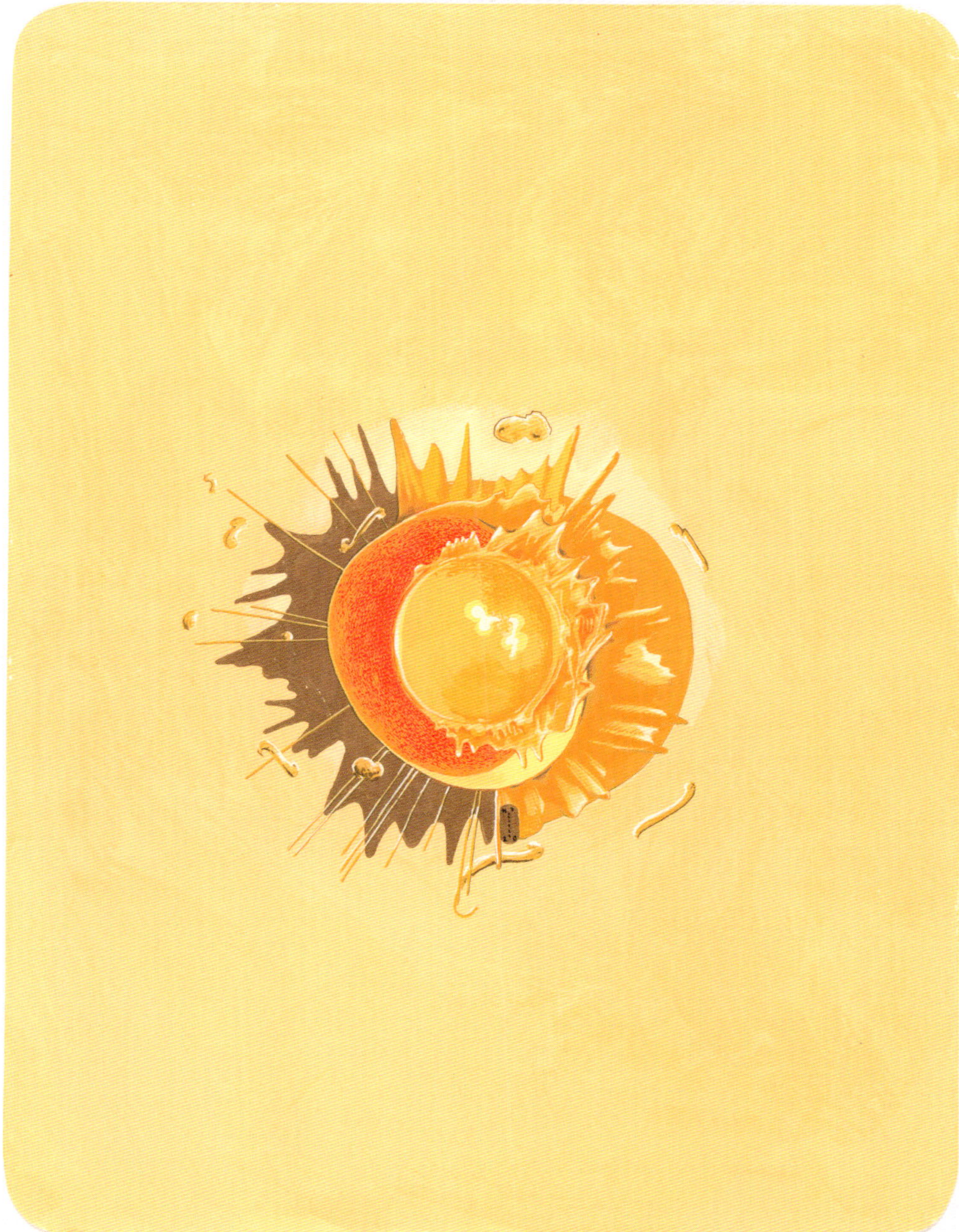

Henry Turner, *The Sun*, 2020, gouache on paper, 435 × 325 mm

eclipses

There's one thing that used to bother me about the amazing relationship between the sun and the moon: If the moon's going around us like it does, then why don't we constantly get in the way? It turns out that it's because the moon's orbit is tilted at 28 degrees to the plane of the equator, which keeps us out of the firing line. Most of the time.

Of course, sometimes things do line up in such a way that the moon finds itself positioned exactly between us and the sun. When this happens, the moon blocks out the sunlight for a short spell. This eerie moment is called a solar eclipse. When it happens the other way around, and the Earth blocks the sunlight from falling on the moon, it is called a lunar eclipse.

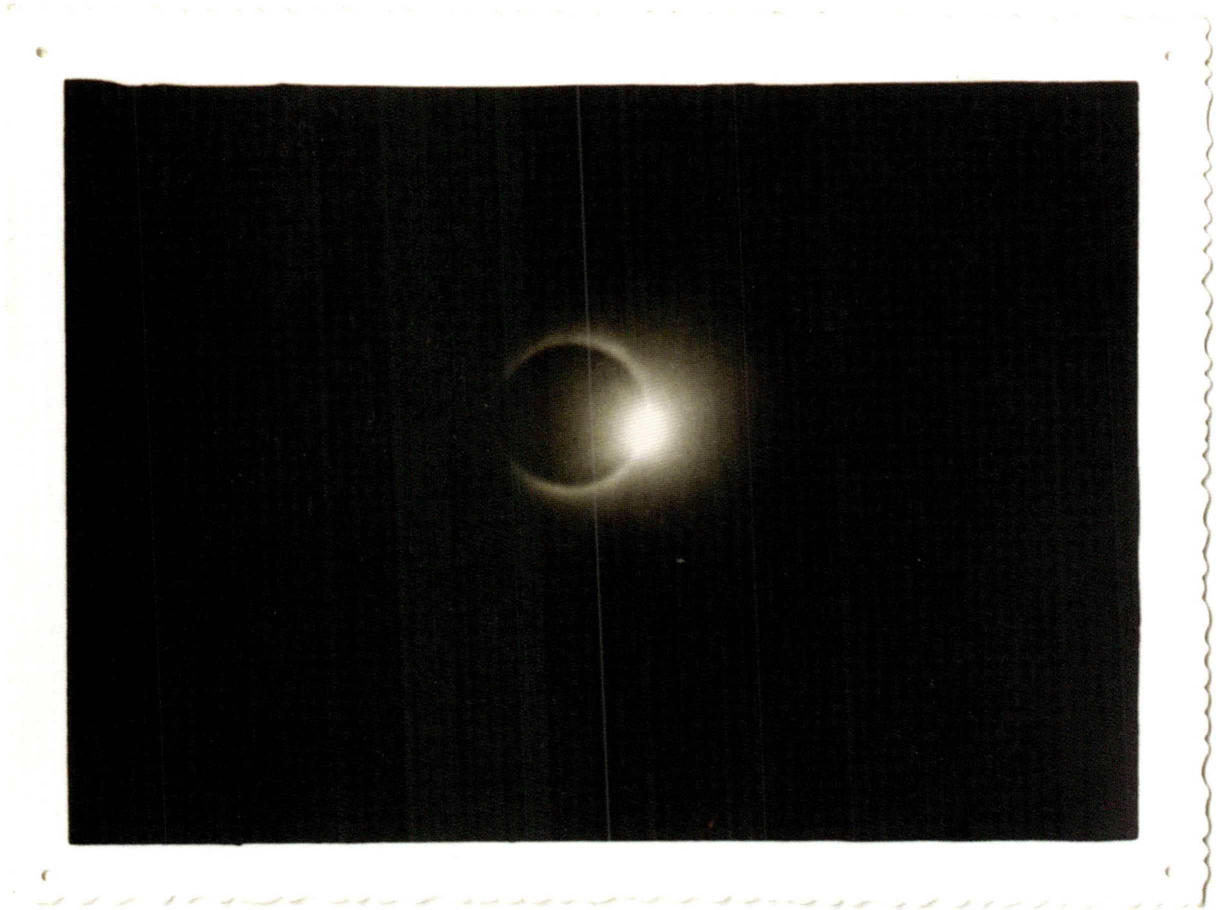

Patrick Pound, *Circle games* (detail from a found collection of photographs), 2020

the sun never sets

I don't really have to tell you that, do I? The sun just stays where it is and the Earth turns around it. At the end of the day, the Earth's horizon comes up to meet the sun, giving the impression that the sun is 'going down'.

The sun hits the atmosphere at a very oblique angle at this point, which gives us the fancy light shows we call 'tropical sunsets' et cetera. Instead of 'sundown' it really should be called 'earthup', and 'sunup' should be called 'earthdown'. But that's not very romantic.

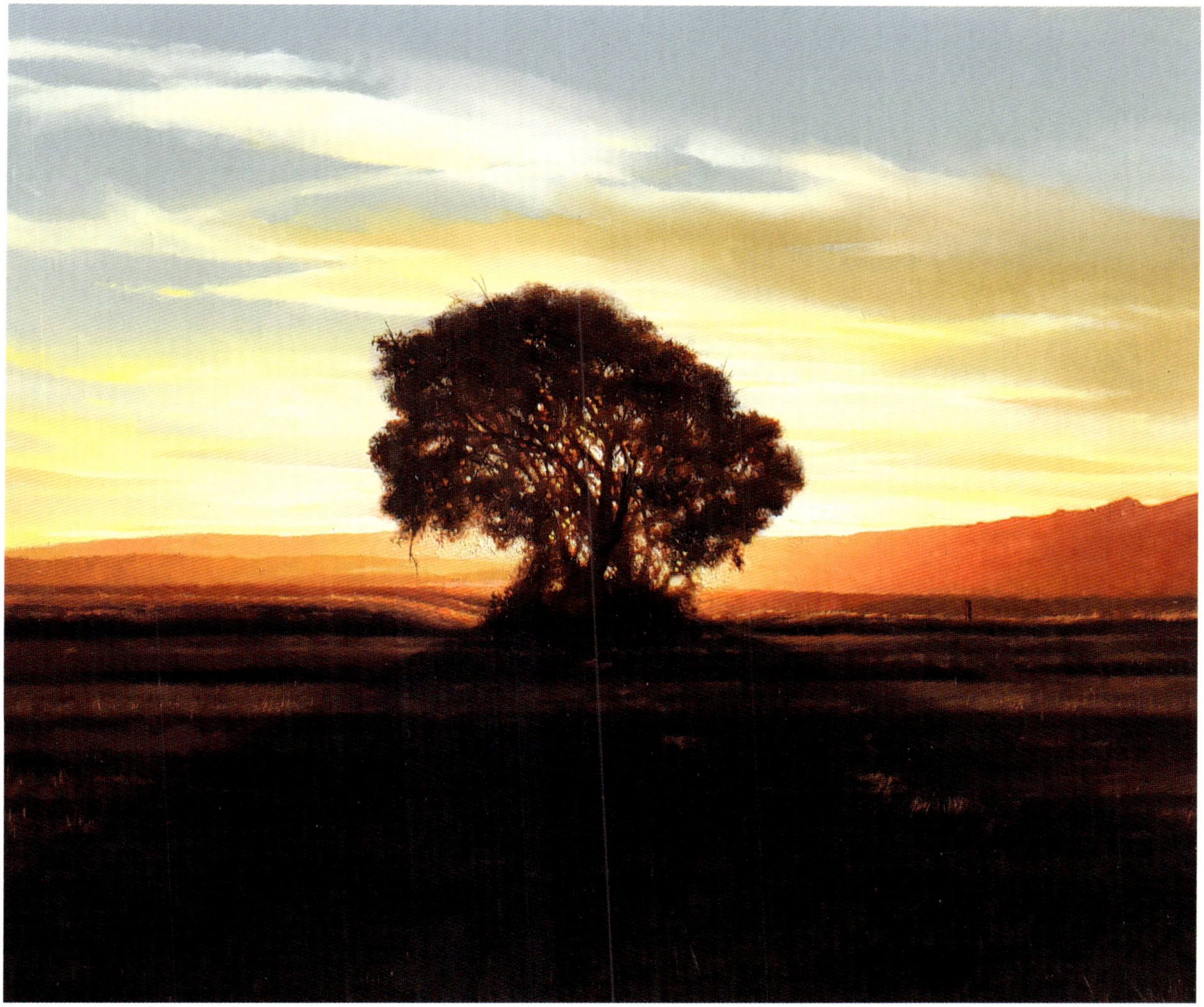

Grahame Sydney, *Sunset, Ida Valley* (detail), 2019, oil on linen, 760 × 1010 mm

going places

Does the sun just sit there, quietly spinning around every 27 Earth days, while we orbit around it? It looks as if it does, doesn't it? (I have no idea how they calculated those 27 Earth days, given that we're all in a bit of a spin one way or another.) However, the sun is on the move, too, taking all its planets with it.

We, and all the other stars and planets, move around the centre of our galaxy. And this huge galaxy of which we're such a tiny bit is called the Milky Way — which is a bit of a lame name in my opinion, not as sci-fi as Andromeda or Centaurus A (but probably better than the Cartwheel Galaxy).

The Milky Way is also on the move, which is something to behold . . . a travelling galaxy! It *seems* to be heading in the direction of the Andromeda Galaxy, which is 'only' 2.5 million light years away. Maybe we're heading towards each other. . . Either way, getting two galaxies together could be a very bad thing. Andromeda is scheduled to collide with the Milky Way in about 6 billion years. Hopefully we'll sort of merge like a zip.

Not very long ago — about 400 years — people did not know all this. They believed that everything revolved around the Earth. It's easy to understand why they would think that: because a) it made them feel very important, and b) it feels so solid down here on the Earth — rock solid. How could we be going anywhere? The sun comes up, the sun goes down, and everything is copacetic in God's Kingdom!

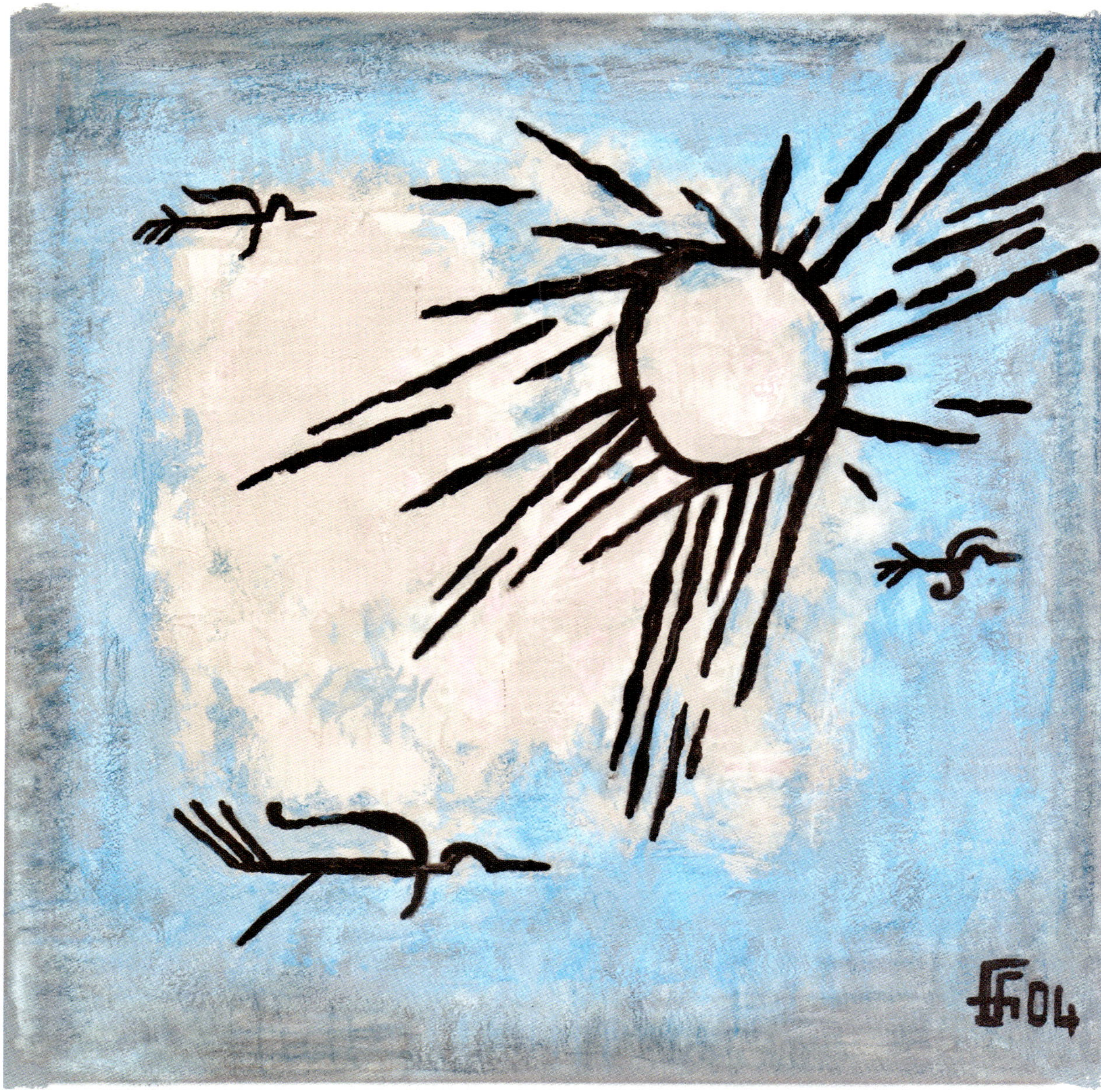

Fane Flaws, *Birds and sun*, 2004

la chiave* in the works

In 1543, an Italian astronomer called Copernicus published a revolutionary book called *On the Revolutions of the Heavenly Bodies*. After thinking about it for 84 years, the Catholic Church decided to ban this blasphemous tome. They believed that the Earth was at the centre of the universe, put there by God, and they didn't want scientists saying that it wasn't.

Roughly about the same time, another creative young scientist —also Italian — called Galileo made himself a telescope. By closely observing the movement of Jupiter's moons, he noticed that everything seemed to be moving around everything else. Galileo began to think that maybe a bit too much was happening out there for us to be the only thing *not* moving!

In 1632 — only 16 years after Copernicus's book was banned — Galileo published his own book. The Church was so fed up with this overwhelming avalanche of common sense that they put him under house arrest. But common sense prevailed: looking through telescopes became a bit of a thing, and the Church had to finally admit that the Earth that they thought God had made wasn't the Centre of the Universe.

* In case you're wondering, 'chiave' is Italian for 'spanner'.

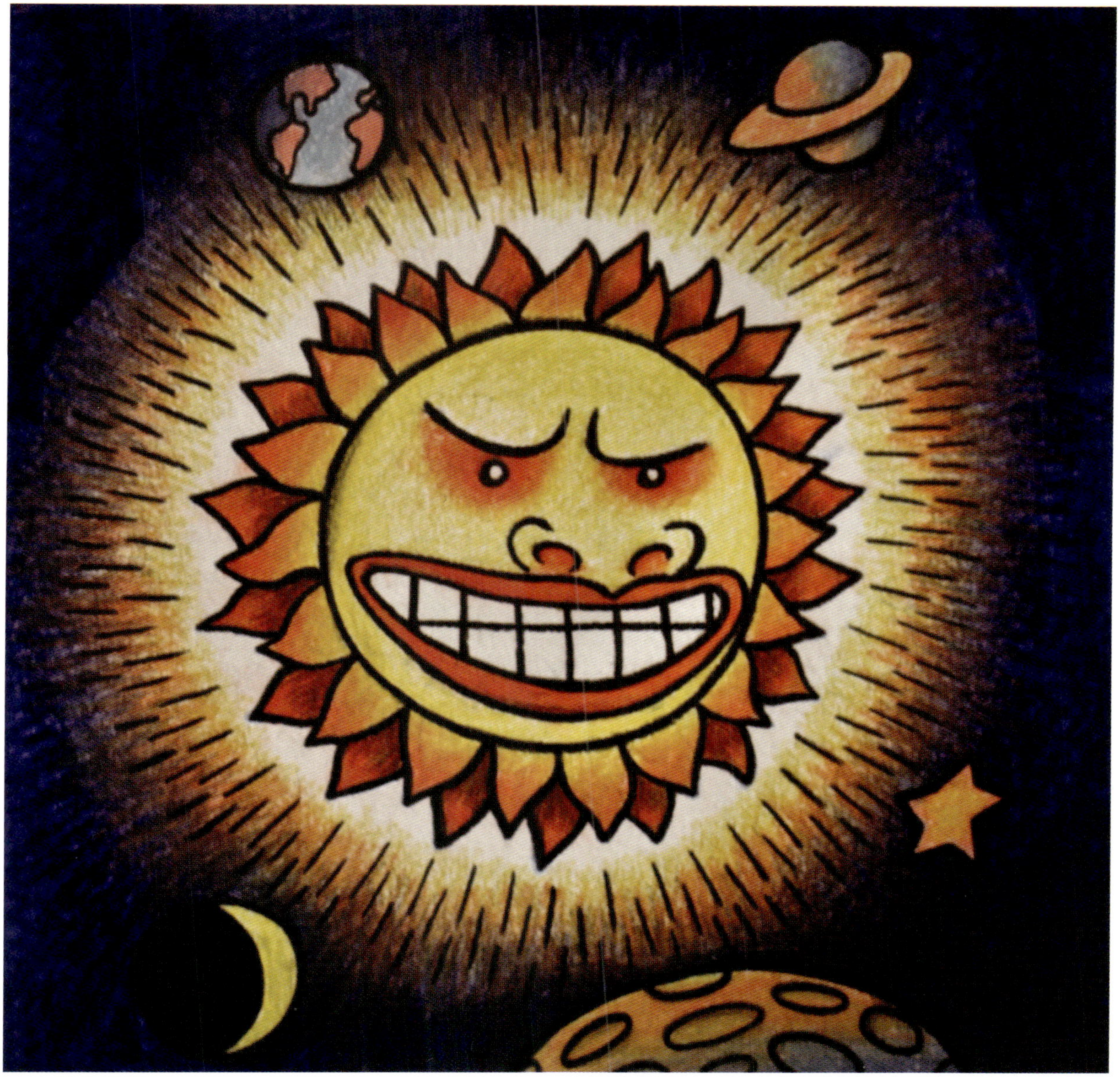

Reg Mombassa, *Angry Sun*, 1998, charcoal and coloured pencil on paper, 380 × 380 mm

light beams

How does the energy of the sun work? Well, basically it works like a totally out-of-control nuclear reactor — roiling and boiling, seething and surging, burping out great explosions of burning plasma millions of miles into space. It's a massive hydrogen furnace.

The energy it releases comes to us in the form of sunlight and heat — lots of heat. They're actually one and the same thing. Fortunately, Earth is in the perfect place to receive all this light and heat — not too close to the sun to get fried, and not too far away to freeze. That's why scientists say we live in the 'Goldilocks zone'.

The light beams that do reach us take eight minutes to get across space, which sounds remarkably quick given that the sun is 150 million kilometres away. If you were to fly there with Air New Zealand, it would take you 19 years. Think about how fast those light beams are travelling . . . as fast as the speed of light, in fact.

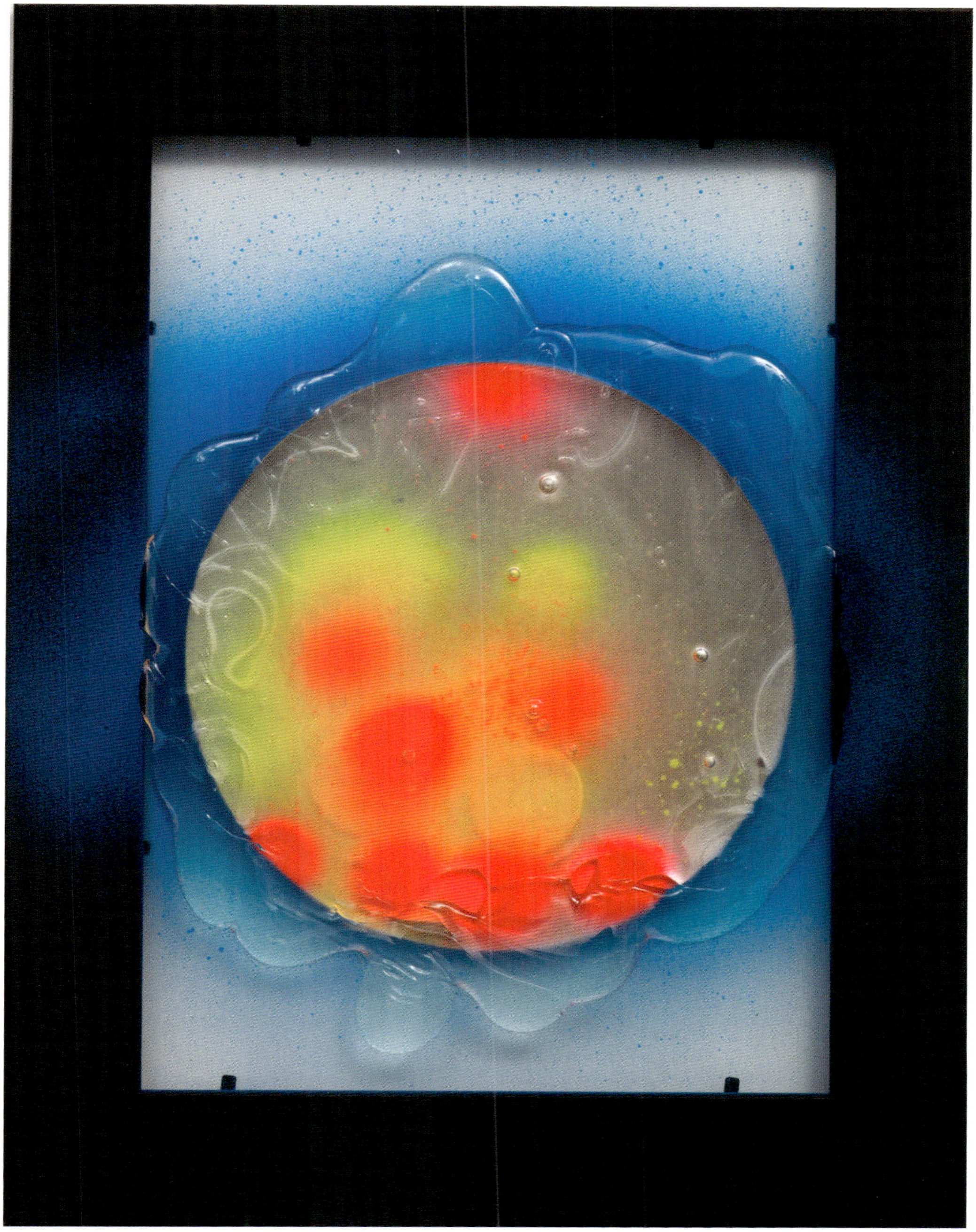

Judy Darragh, *Globby Sun*, 2017, silicone, acrylic paint, mirror, wooden frame, glass, 200 × 300 mm

light feed

Where does all the sun's energy go to when it gets here? How do we benefit from it? That's a big question — very big, and very important, because we wouldn't exist without this energy.

Sunlight powers all the food production on Earth. Plants absorb the sunlight using an amazing process called photosynthesis, which gives plants the fuel they need to grow using water and air. If this were a botany lesson I'd explain how that works, but for now you'll just have to take my word for it.

We eat the plants, or animals eat the plants and we eat the animals and the plants. This is how the sun's energy feeds us.

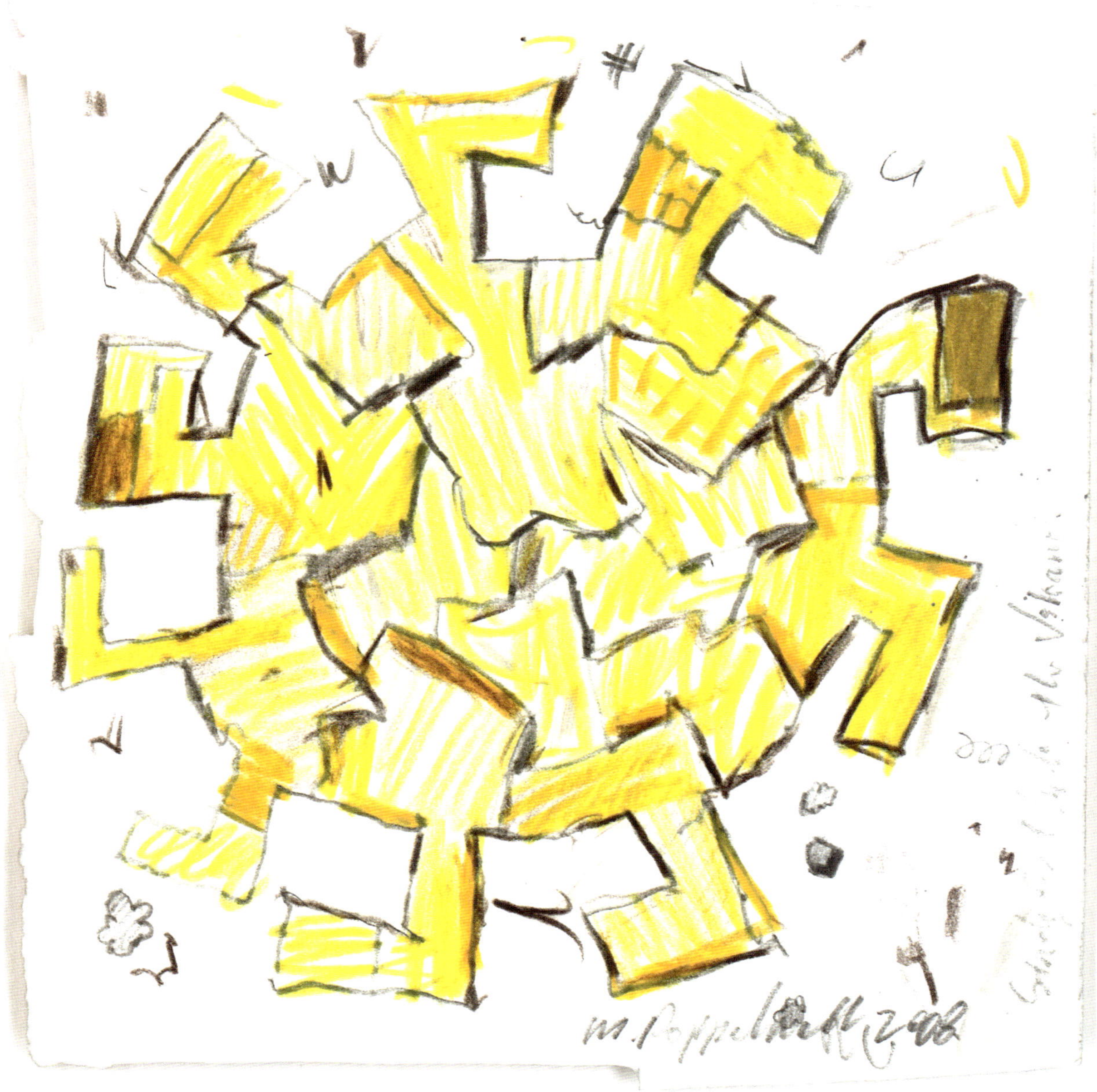

Martin Poppelwell, *Sun study (under the volcano)*, 2008, pencil on paper, 130 × 130 mm

digging it

The incredible energy that comes from the sun is even stored in vegetation and organisms that rotted down many, many years ago and gradually became covered with mountains and oceans. All these years later we dig it up in the form of coal, oil and gas. We then burn the coal, oil and gas, and it releases all the stored energy. That's the heat of the sun you're feeling there in front of the fire!

Nowadays all this mining and burning is proving to be bad for our planet. So, we're working hard to find other ways of getting the energy from the sun directly, using much cleaner processes like solar panels and photovoltaic cells. It won't be long before we look back on all this coal and oil drilling, digging and burning as an incredibly dirty part of our history. All that smoke, smog, and soot? It's starting to look a bit dated already!

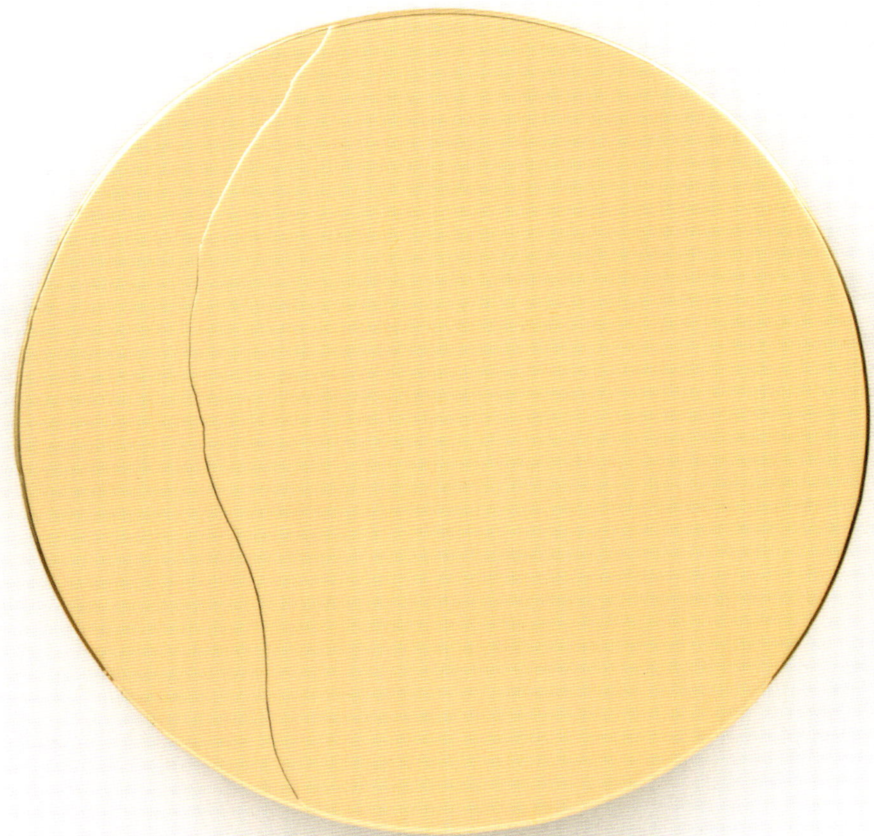

Max Gimblett, *As above so below*, 2020, gesso, resin and gold leaf on wood panel, 381 × 381 × 50 mm

the ubiquitous electron

So far we haven't found anything that travels faster than light, though scientists *think* they might have found an electron that travels so fast that it gets where it's going before it leaves. Some scientific wag has suggested that there is only one electron in the entire universe, and it just happens to be everywhere at once.

This fascinating theory is proving to be a bit hard to verify, but given how wacky the world is at sub-atomic levels I'm just about ready to believe it. (I didn't mean to delve into particle physics at this point, but I couldn't stop myself.)

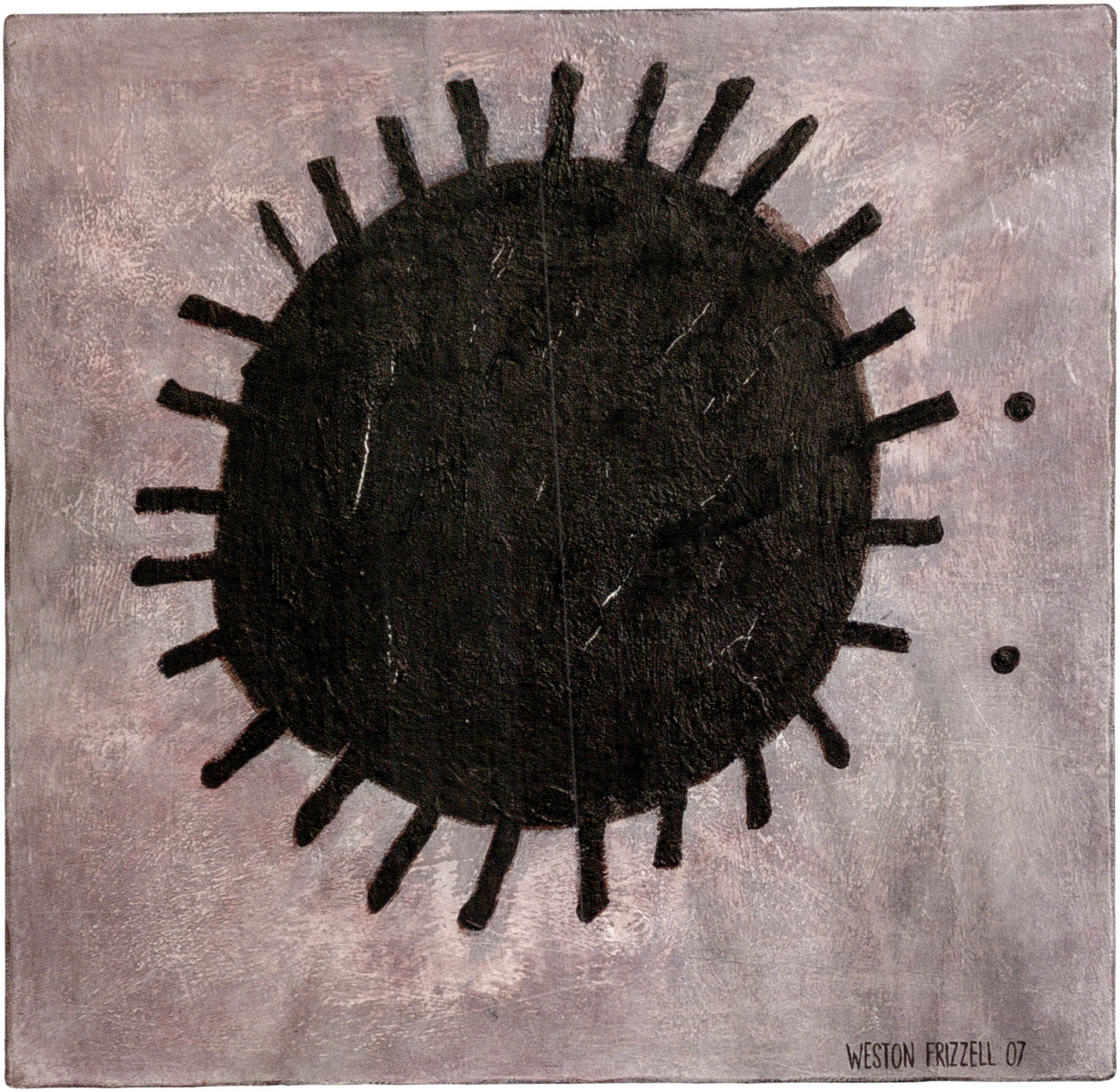

Weston Frizzell, *Black Hole Sun*, 2007, enamel and alkyd on hardboard, 720 × 720 mm

for every season

Why doesn't the sun do its job quite so well in winter? Well, that's a fascinating story.

Very, very early in its life, the Earth got an almighty smack in the head from a massive meteorite. This knocked the Earth out of kilter and put it on the tilt that it's still on today. (By the way, scientists believe that all the rocks and rubble that the collision knocked off the Earth were gathered up by gravity and eventually became our moon. A happy accident!)

Because of the Earth's elliptical orbit, this tilt means that half the year the northern hemisphere gets the sun shining on it directly, and for the other half the southern hemisphere gets it.

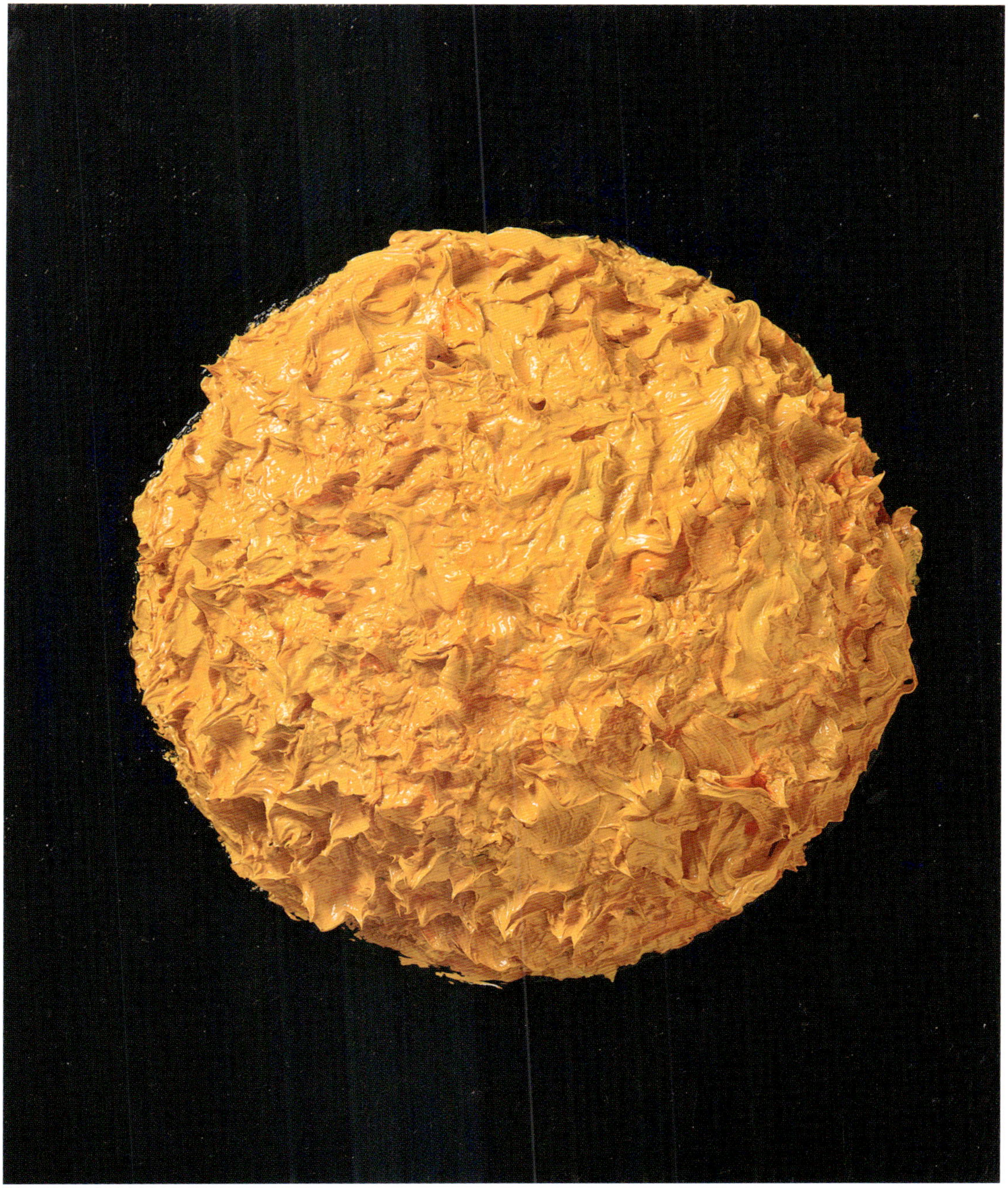

Karl Maughan, *Painted sun*, 2021, oil on canvas

and here's how it works

Imagine that you are holding an orange out to one side of your bedside lamp, and you tilt it towards the light. You would see most of the top of the orange all lit up. If you then moved the orange to the other side of the light, still holding it on the same tilt, most of the bottom of the orange would be lit up. Now imagine the orange spinning around at the same time, and you get night and day. Spring and autumn happen in the time it takes for the orange to get from one side of the lamp to the other. Remember, it takes the Earth a year to orbit the sun, which means six months to get from one side to the other.

If there was no tilt on the Earth, there'd be no seasons and no moon. And if we had neither of those things, we wouldn't be here talking about oranges and bedside lamps! It is thought that the extraordinary and dynamic conditions created by this highly unlikely conflation of events provided the unique environment that allowed life to evolve.

If you add the fact that that our moon's distance from the Earth makes it appear to be the exact same size as the sun, meaning that we're possibly the only place in the universe where a perfect eclipse can occur, you start to get an idea of just how special we are.

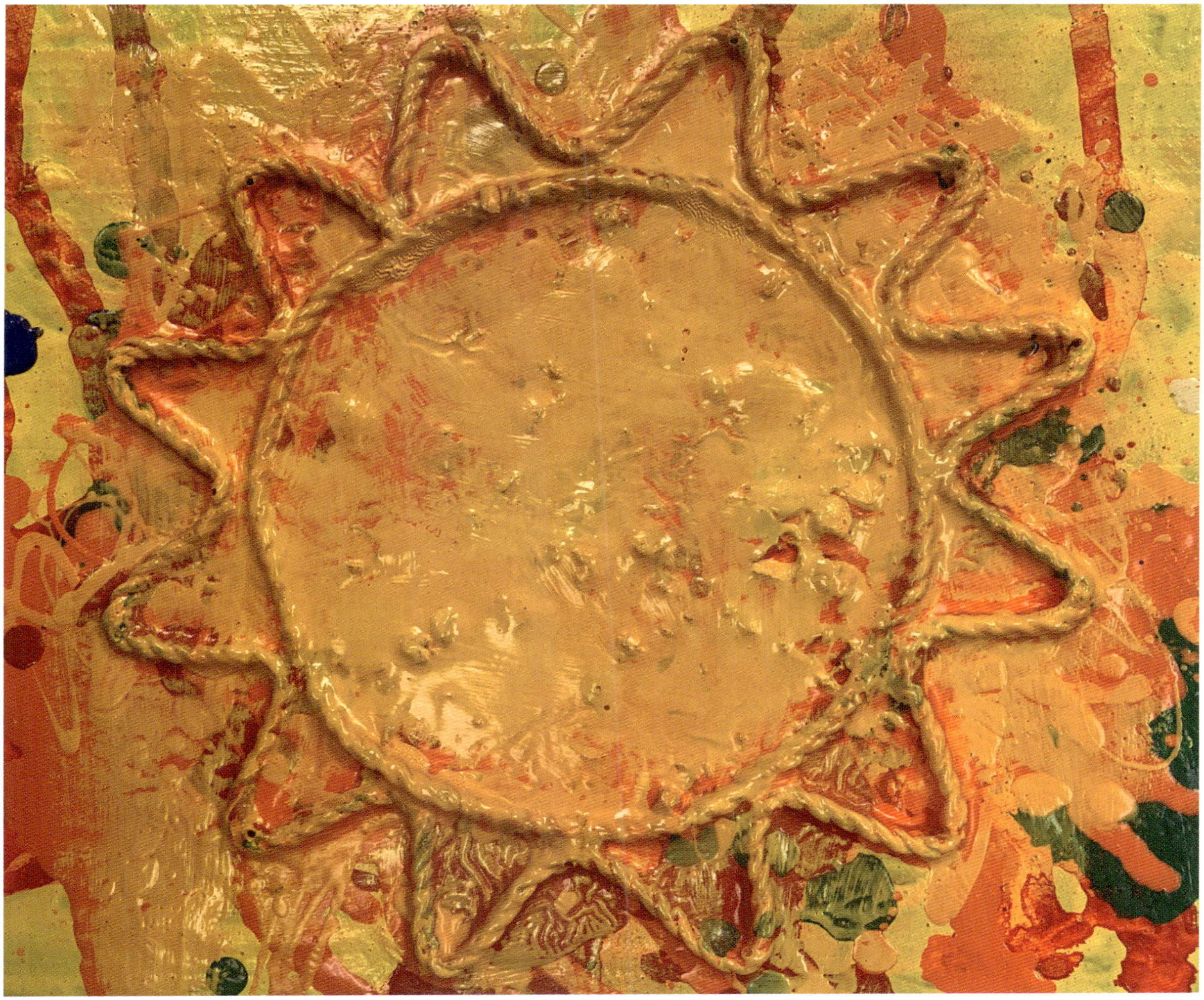

Dick Frizzell, *Jaune Soleil*, 2021, mixed medium on board

a million points of light

Like most people, I look up at the stars in the night sky and picture them all as single points of light scintillating away in the heavens. But it's way more complicated than that.

Apparently, at least 70 per cent of those singular-looking pinpricks of light are actually made up of more than one star. There can be up to six stars contributing to a single point of starlight! These amazing groupings are called multiple star systems, which is a pretty straightforward description of an insanely complicated phenomena: six star systems doing a gravitationally shared dosey-doe way up there in space.

This is too much for me to get my head around, so I'll just talk about two stars doing that dance. Two stars together are called a binary star system. This means that two entire star (or solar) systems, each with their own planets, moons, asteroids and whatever else they're dragging around, are quietly going about their own orbiting business, while *at the same time* somehow managing to orbit each other.

And there's more — this whole complex arrangement is also continuing to orbit around the centre of the galaxy. That's a lot of orbiting. It'd be a mistake to imagine this looking like some cosmic Catherine Wheel. This is space, remember, and everything is vast — even the vastness is vast.

We now think that our own solar system was once part of a binary system. We think this because there are some gravitational and orbital quirks in our planetary system that can't be explained. It's a bit like people who have lost arms or legs but feel as if they are still there!

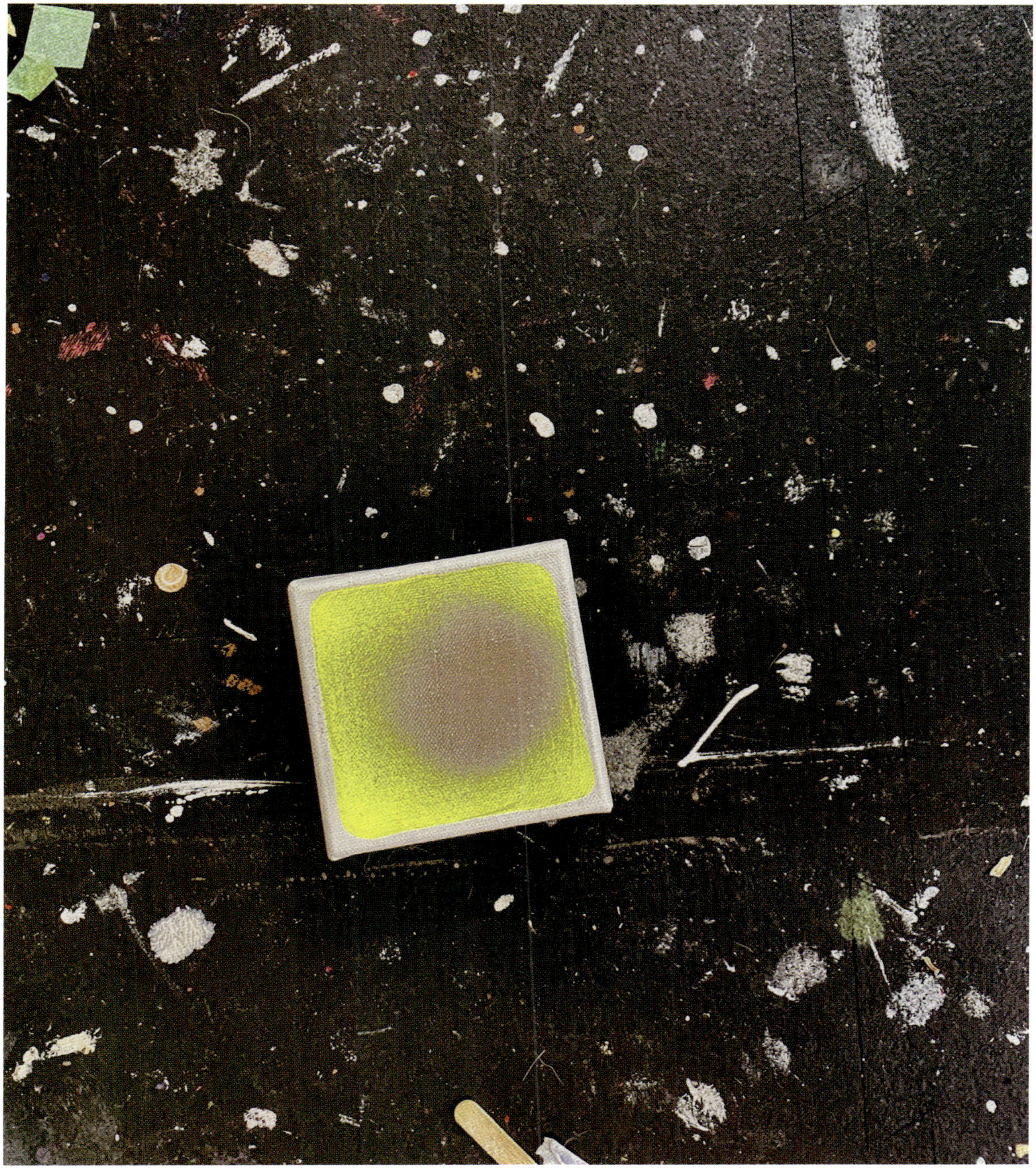

John Reynolds, *The Icarus Files*, 2021, studio floor, metallic paint and acrylic on canvas

keep on burning

What keeps the sun burning? The simple answer is that the sun burns itself. There is no one up there to throw another log on the fire — it's all in there already. The good news is that there are enough logs up there in the sun to last for billions of years. The bad news is that eventually the logs will run out. But don't worry about this — by the time the sun does flare out we'll be long gone, hopefully to a younger solar system than ours. (I'll say it again: DON'T WORRY ABOUT IT!)

From our new home we'll be able to look back and watch our sun go *kablooey!*, burning up most of the solar system in the process. There'll be no going back. But don't let this keep you awake at night. This spectacular event is 7.5 billion years away.

And don't lie there staring at the ceiling trying to get your head around that mind-boggling figure either. I've never come across a convincing way of picturing just how long that is, but this might ease your concern: we have been on Earth for 6 million years, and it looks as if we're good for at least 7000 million (that's 7 billion) more. We should have an exit plan figured out by then.

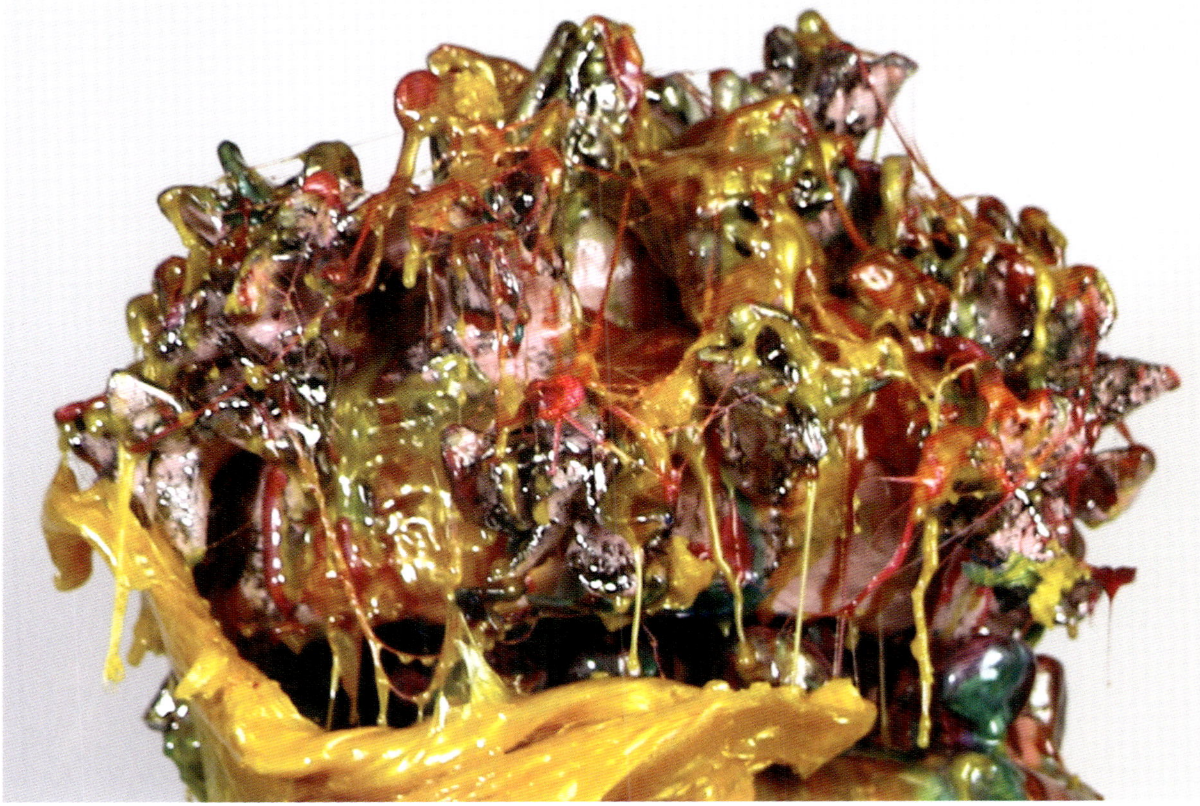

Virginia Leonard, *And the dawn comes up like thunder outer China ‘cross the bay*, 2020, clay, lustre and resin, 750 × 500 × 500 mm, courtesy of Gow Langsford Gallery

the origin story

OK. Where did our sun-star come from? Or, more to the point, how come we're even here to ask the question? What did the singer Joni Mitchell mean when she sang about being stardust and golden?

It's a long story, but basically the universe seems to have evolved out of a mysterious moment called the Big Bang. We're still a bit hazy on this one. I find myself wondering how could there even be a bang if there was no air to carry the sound waves or ears to hear them?

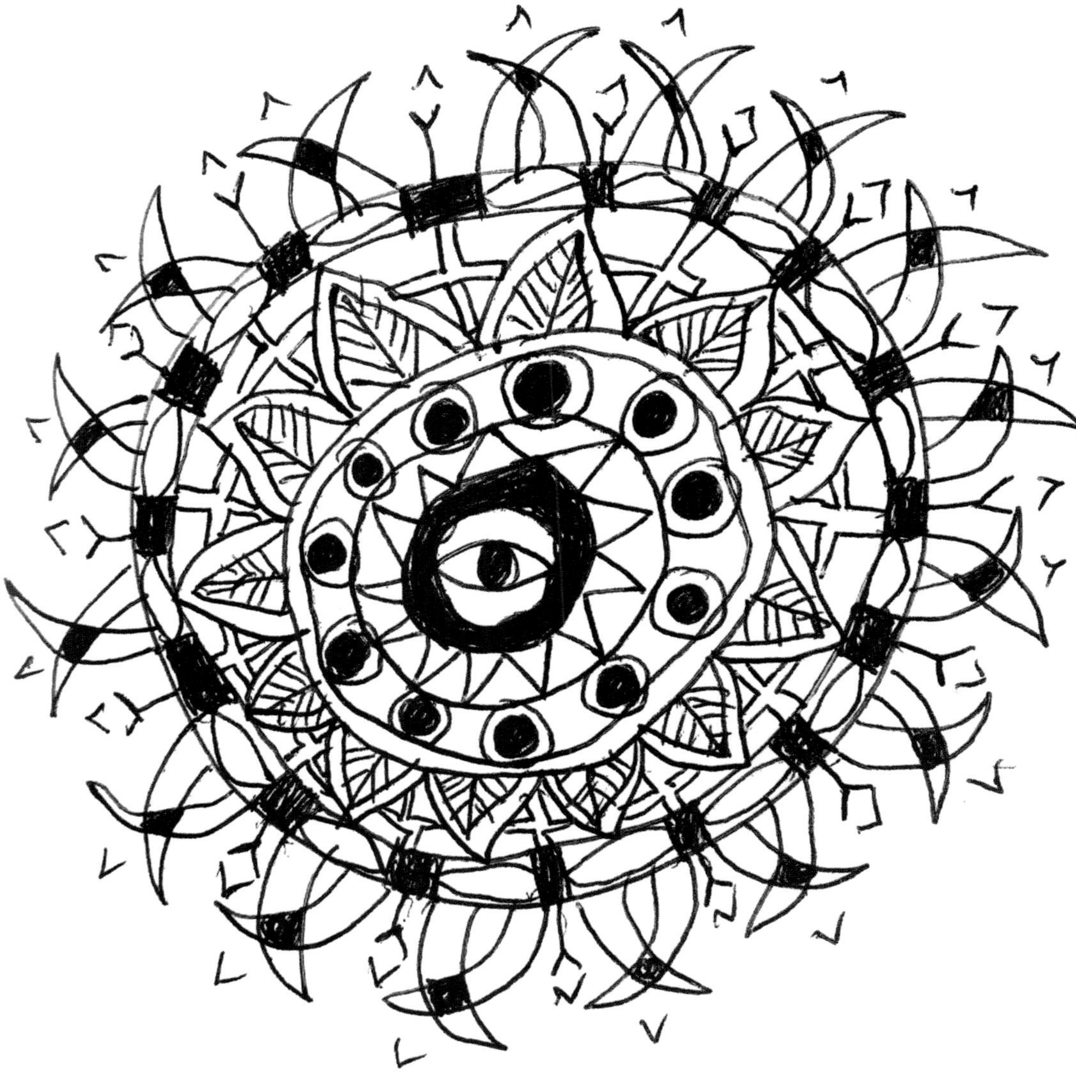

John Pule, *My star*, 2021, ink on paper

free lunch

Anyway, apparently everything needed to make this enormous thing called the universe kind of popped out of nothing in one almighty hot and concentrated split-second of infinitesimal magic. (Not really magic, of course, but physics — though I'm inclining towards magic given the weirdness of the concept.)

This unbelievably hot, dense dot of plasma quickly (oh, so quickly) spread out, cooling down as it went, like hot soup cooling down as it is spread out across a very large plate. Then our other mysterious friend, gravity, emerged from the soup and started its heroic work, clumping all the stardust together into spinning masses that formed the very first stars.

A bit like the way a pearl in an oyster forms around a tiny impurity, the mass of these first stars formed around little irregularities in the formless force-fields swirling about in the soup. In fact, the soup was nothing *but* force-fields up to this point.

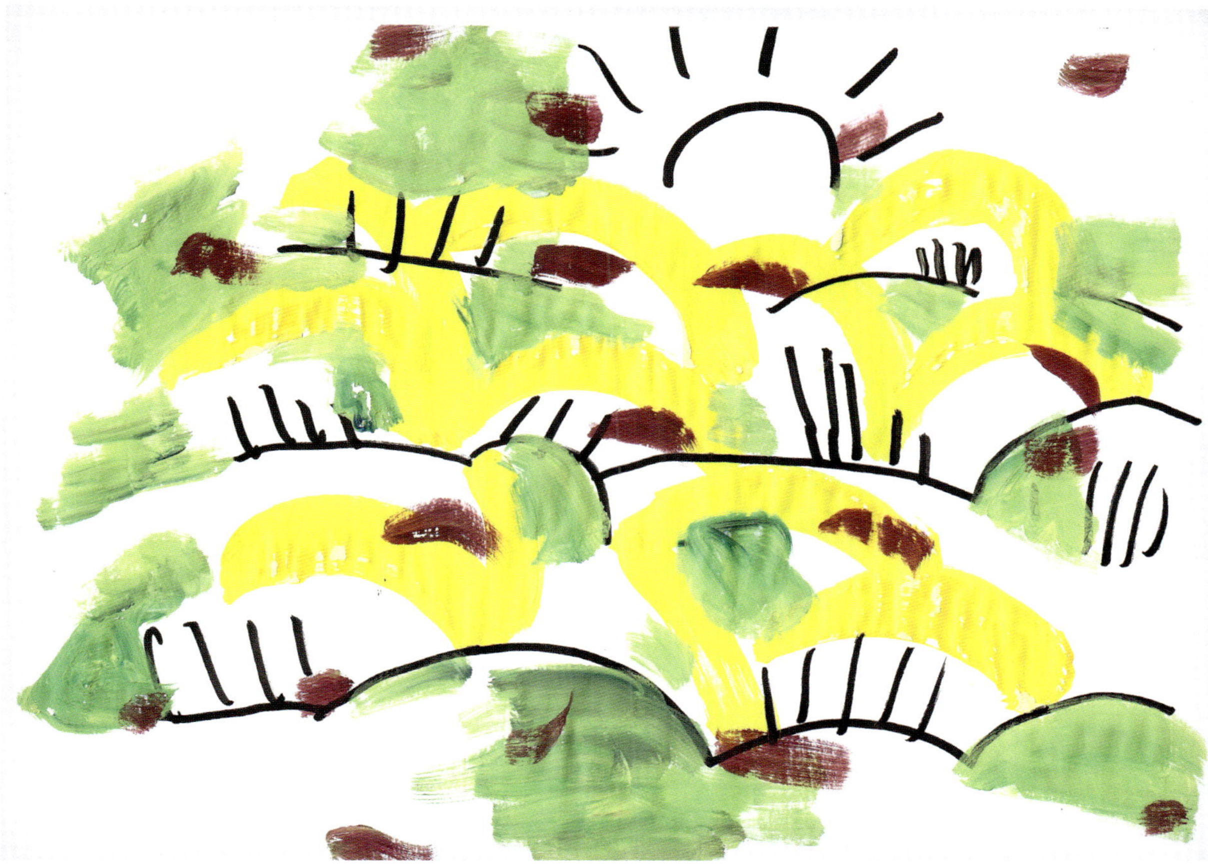

Phil King, *Winter Harvest*, 2020, acrylic and marker on paper, 210 × 297 mm

fairy dust

This is the *really* amazing bit. Very early on in the expansion of the young universe, all the complex chemicals and compounds necessary to make anything as complicated as life didn't exist. The makings did, but not the actual stuff. The first atoms to make an appearance were the simplest, like hydrogen and helium. That's why the first generation of stars were made of clustered clouds of hot hydrogen atoms. These clusters got denser and denser as more and more atoms came to the party, creating the perfect conditions for new compounds to be created deep in the core of these insanely hot furnaces.

Then, as things usually do when they get too hot and heavy, these stars themselves exploded. Their magic fairy dust got scattered all through space, where it also started spinning together and clumping up again into the atoms and molecules needed to make life.

So, even though it's a great leap from this 'stardust' to us, Joni was basically right. And it's a lovely idea, isn't it?

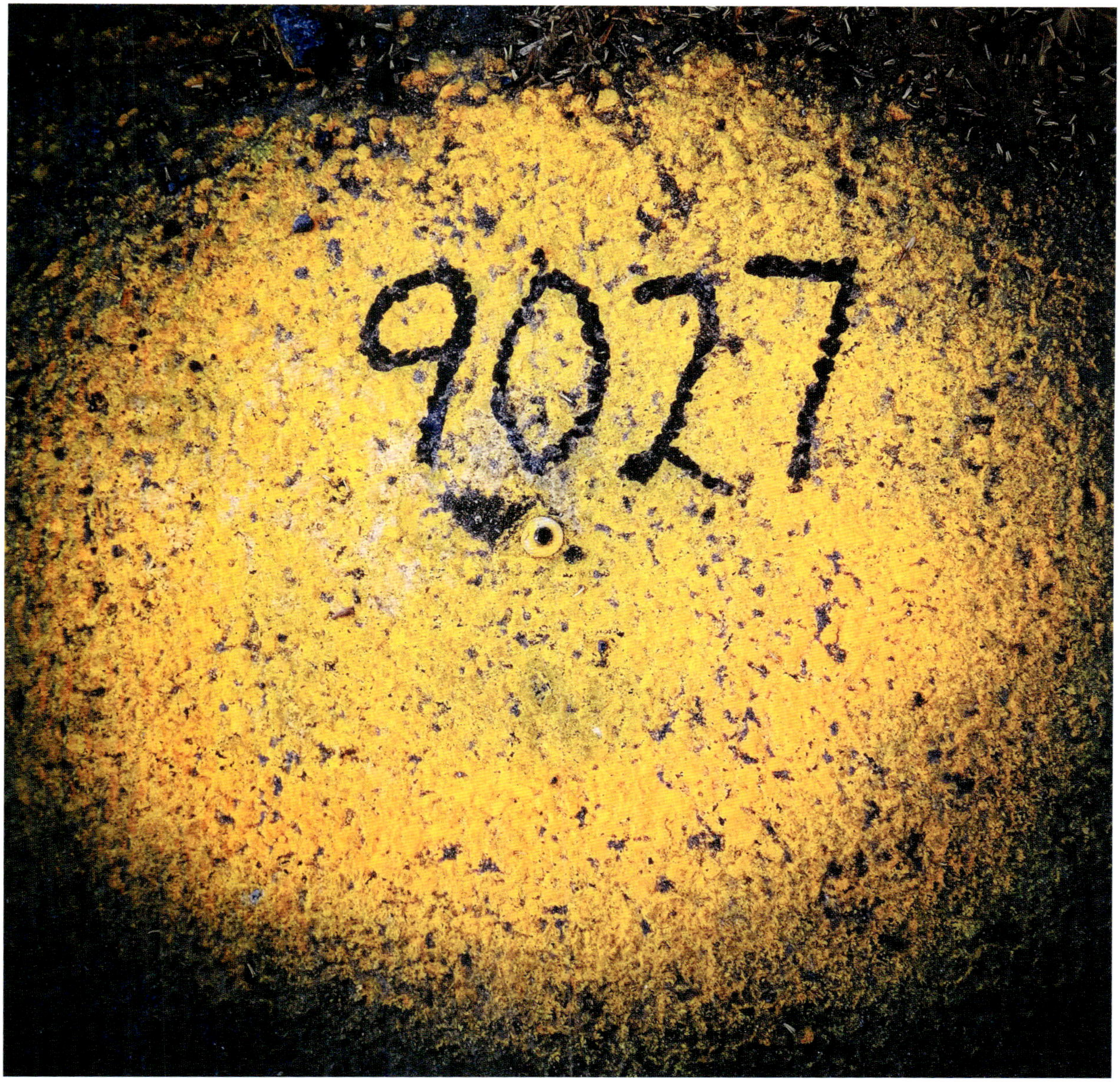

Thomas Newman Pound, *9027*, 2018, found image

big, but not big enough

It's taken about 4 billion years to get from all that stardust to where we are now, and that's about half the time we reckon we've got before our sun gives up *its* stardust. Obviously a lot can happen in a few billion years.

Oops — seems I might have got that last bit wrong. Our sun isn't big enough to go the full supernova; instead it will flame out, shed all its ashy outer layers into space and then shrink down to what we call a white dwarf. This white dwarf will be about as big as our Earth (which won't exist anymore at this point, remember?) but insanely denser — which makes sense given how much sun will have collapsed into it.

Their massive density gives white dwarfs incredible gravitational pull, but not as much pull as black holes, which are so dense that they even trap light waves. That's why they're called black holes. White dwarfs shine very, er, whitely, but in reality they're just glowing like embers. The fuel is all gone, and eventually they quietly fade away.

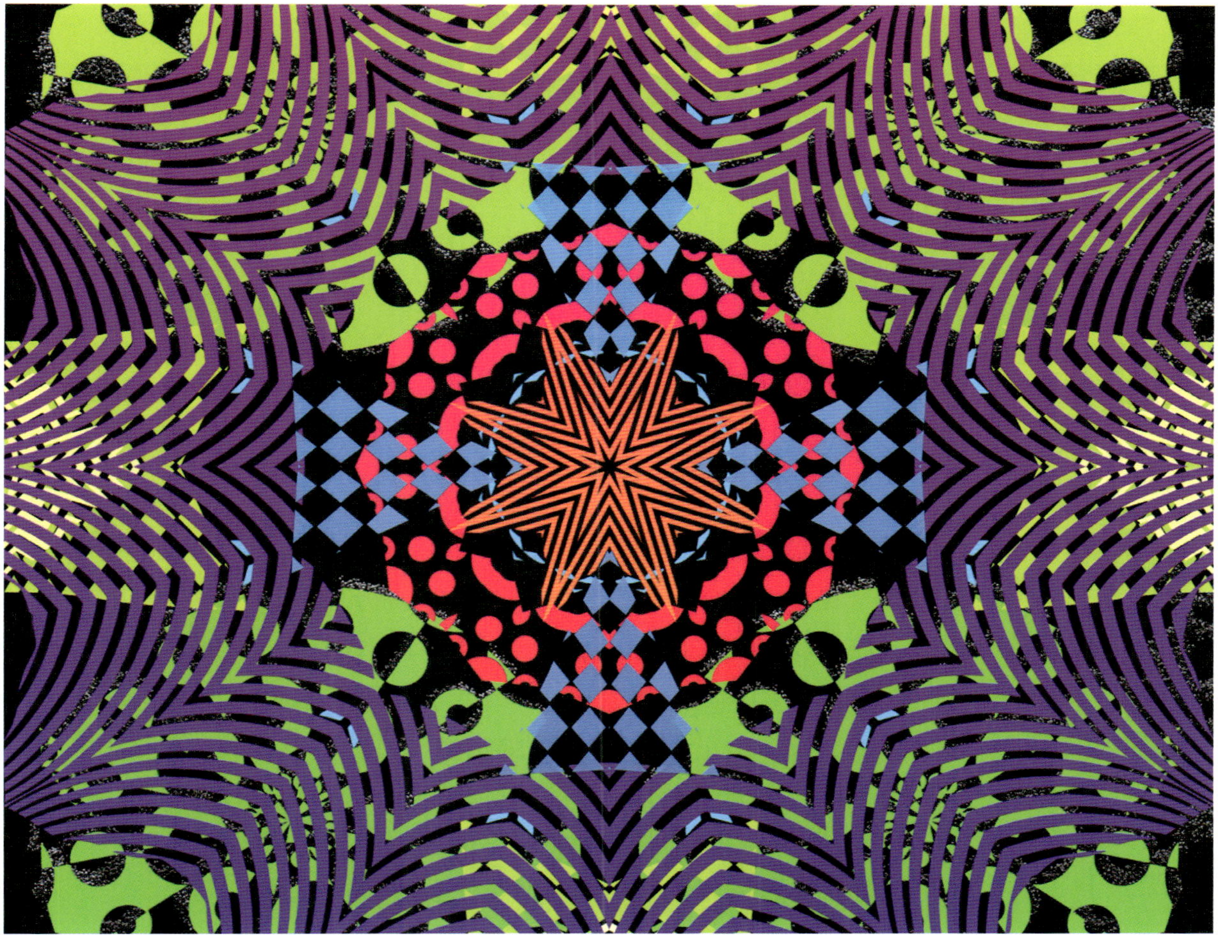

Reuben Paterson, *Te Maiea* (still detail), 2020, digital animation (4:30-minute loop)

space and time

When I was little I used to lie awake in bed trying to get my head around the idea of infinity. How could the universe have no end? No boundary? What if you *could* get to the end of the universe? Couldn't you just cross over? While you're there? Wouldn't even nothing be *something*? I used to literally mind-boggle myself to sleep.

This question bothered me for years, until I got old enough to read some real science (not just Buck Rogers and Flash Gordon comics). I then found that space and time are the same thing. You can't get ahead of time — you can't just step into tomorrow anymore than you can step beyond the edge of the universe.

The universe — and everything in it — is always expanding into tomorrow. That's a bit hard to believe if you overthink it, but it's quite straightforward really. When you blow up a balloon, it expands into the party room. When the universe expands, it expands into tomorrow.

What a scene, eh? More magic than magic. Magic, mysterious and beautiful. And here we are. On the third rock from the sun, figuring it out.

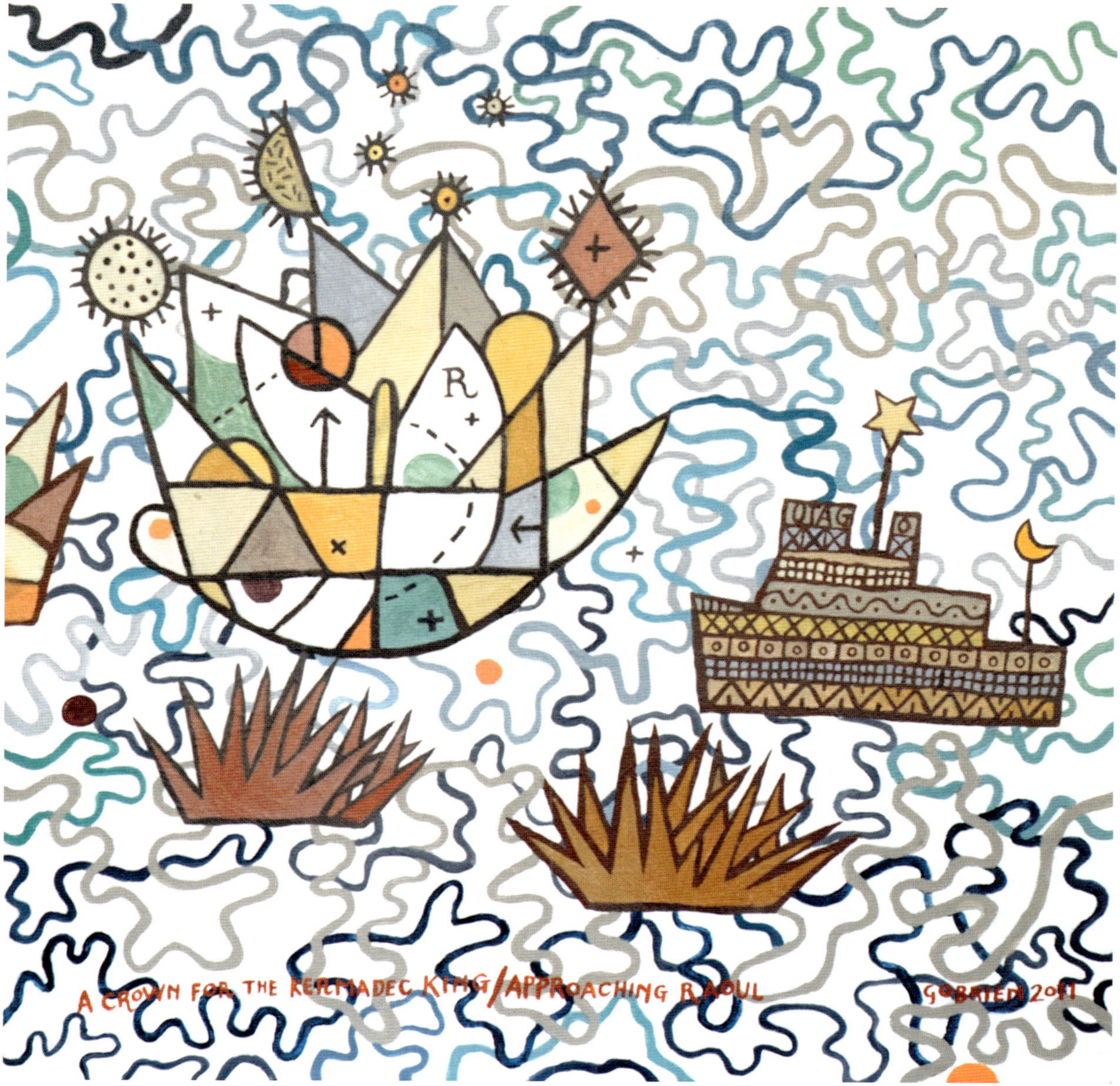

Gregory O'Brien, *A crown for the Kermedec king/Approaching Raoul*, 2011, acrylic on hardboard, approx. 750 × 750 mm, collection of the New Zealand Maritime Museum Hui te Ananui a Tangaroa

lessons to be learnt

This is what we've figured out so far — being very careful to remember that we know how the universe works after it came into existence, but not how it came *into* existence.

We understand that in the split, split, split, split, split, split, split, split, split, split second that happened after the initial mysterious nothing, everything suddenly existed: space, time and everything else appeared in its infinite entirety. Unfortunately, the conditions of the universe in the infinitesimal moment when it came into being were so bizarre that we don't have the physics to describe it.

(I keep using the word 'infinite' here because it's beyond me to conceive of anything else. I'm happy to acknowledge that it's a bit inadequate when you consider new speculation about what might exist beyond the edge of the observable universe. But let's not go there — not in this book, anyway.)

So space (the same thing as time, remember) was also infinite the moment it happened. *But* — and here's the tricky bit — as well as being infinite, it was also dense. Oh boy, was it dense. No space in the space. It's really beyond what we can imagine: everything we understand about the infinite universe was right there in that moment. *Everything*. Not in the form we see it today — and I've explained how that evolved — but all the ingredients were there in that unimaginably hot, dense, infinite moment.

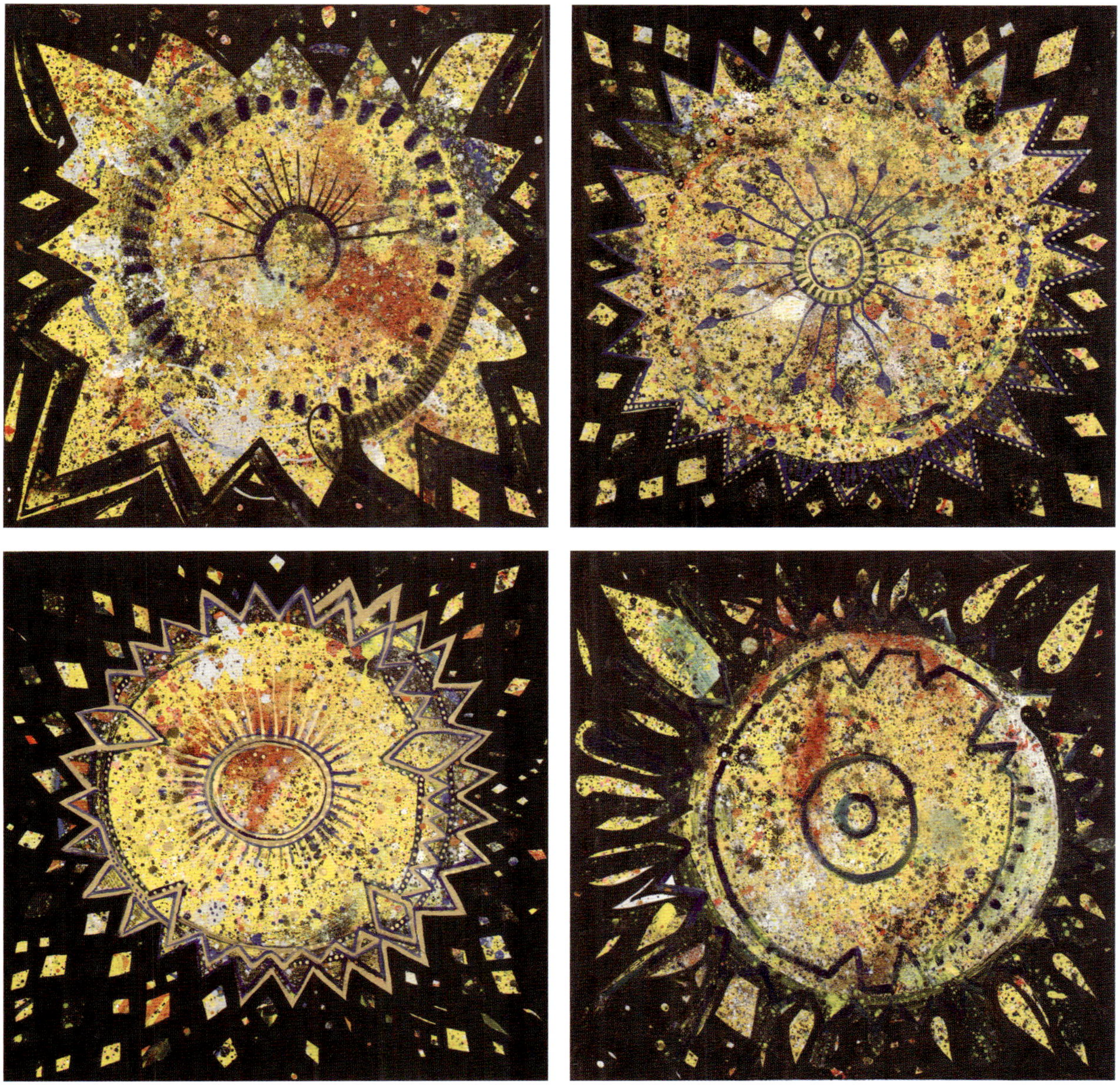

Richard Boyd-Dunlop, *Solar 1*, *2*, *3*, *4*, 2021, acrylic and glitter on canvas, 800 × 800 mm

things not adding up

Strangely enough, despite that mysterious first split-second of the universe coming into being, we have a pretty good handle on how it works. Although the universe appears to be mostly space, we've been able to calculate how much matter there is in it (don't ask me how). We thought this was quite clever until we found that all the actual stuff out there — all the stars, planets, moons, dust clouds, et cetera that we measured — accounts for what the laws of physics quite clearly tell us is only 5 per cent of the actual mass in the universe. Where's the remaining 95 per cent?

Well, we're not sure. We've kind of made up these two things called dark energy and dark matter. Dark matter is supposed to account for the fact that we know that *something's* going on out there when we observe the gravitational behaviour of galaxies, but we can't actually *see* anything. It's a bit like the wind. You can't actually see the wind, but you can see what it does. You know it's there because of the evidence of your senses, but you can't actually see it.

Dark energy, on the other hand, seems to be a weird sort of anti-gravity that we've dreamed up to account for the observable fact that something is pushing everything in the universe further apart.

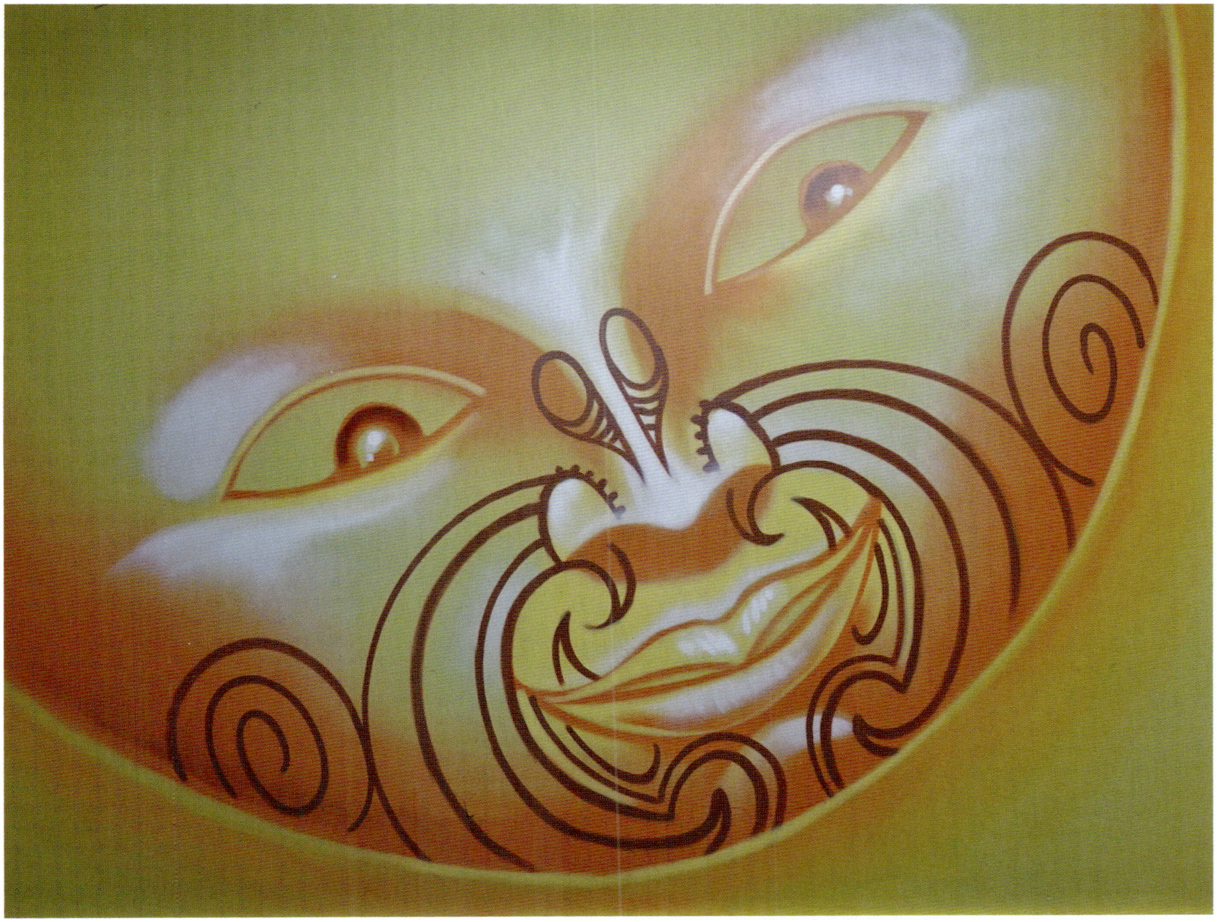

Darryl Thomson, *Tamanui-te-rā*, 2014, mural

something's lurking out there

Say we accept the existence of dark matter and dark energy — but we still have to keep in mind that space is still mostly space. In fact, space is 99.9999999 per cent space. Seriously! Despite those lovely drawings of little balls whizzing around each other, what an atom really amounts to is not much more than an electromagnetic thought bubble with a couple of vague ideas circulating in it.

It's hard to get across mind-boggling things like the amount of space in an atom. By now, you'll be wondering how anything as solid as a rock could ever exist with all this space in it; but that's how atoms, molecules (clumps of atoms) and their interdependent force-fields work. The atoms in a rock just hold on to each other a bit tighter than, say, the atoms in a soap bubble.

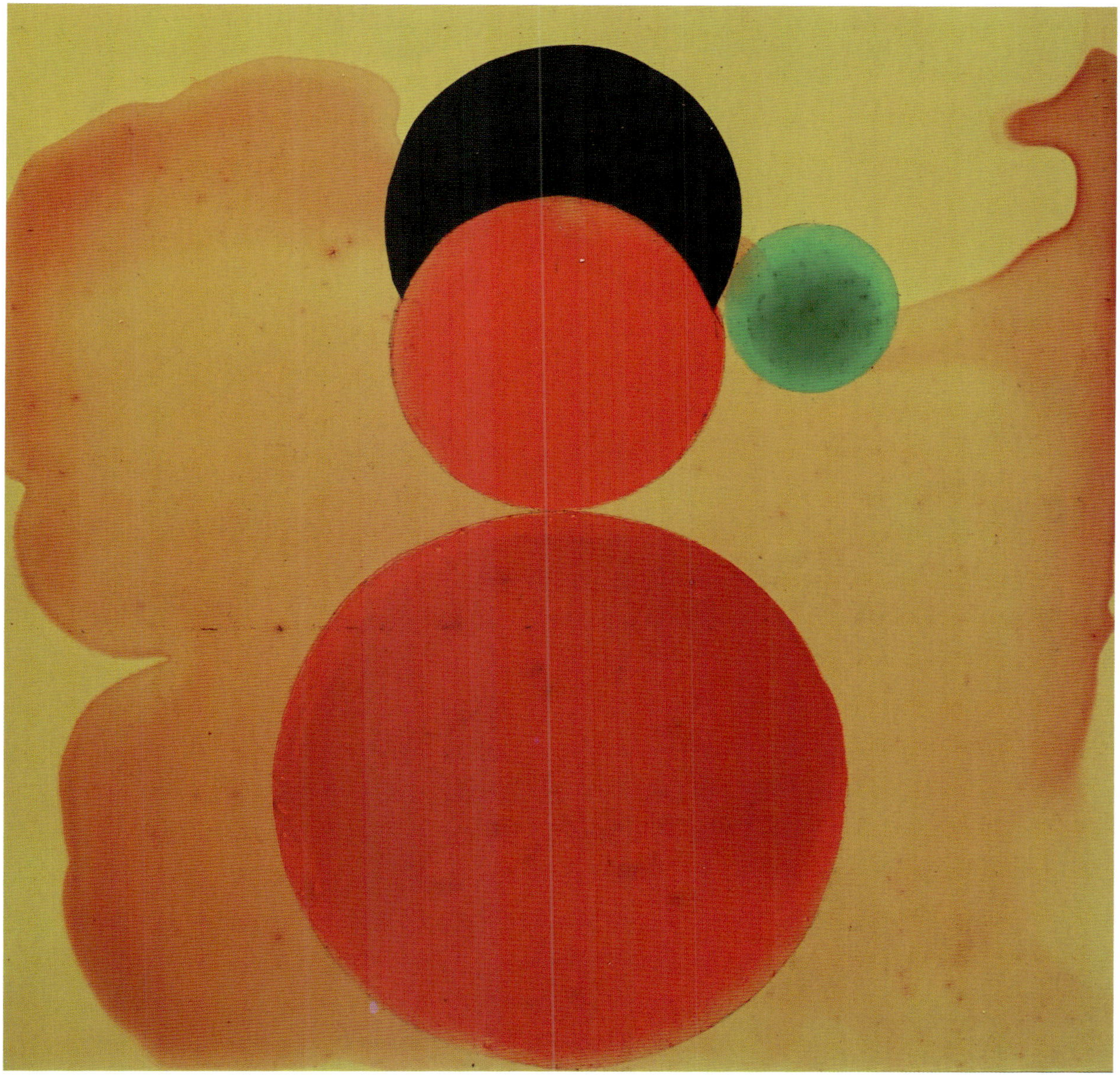

Denys Watkins, *Nickell Atoll*, 2014, acrylic on ceramic panel, 200 × 200 mm

up front and centre

Here's another fabulous thing about that moment at the beginning of time. People still think that the Big Bang must've been like a bomb exploding and blowing outwards from a central point, as bombs tend to do. So they imagine that the universe must have a centre somewhere. Well, it doesn't. The universe is all 'centre'. It's just one big centre, getting bigger every day.

The centre didn't expand outwards; it just got bigger. Everything in the universe expands out from everything else, in every direction, at the same rate, at the same time. It's a bit like a loaf of fruit bread in the oven with all the raisins moving away from each other as it rises. Anywhere you choose to point at is the centre. Wherever you stood in the universe, you'd notice everything expanding outwards from you.

I told you it was weird!

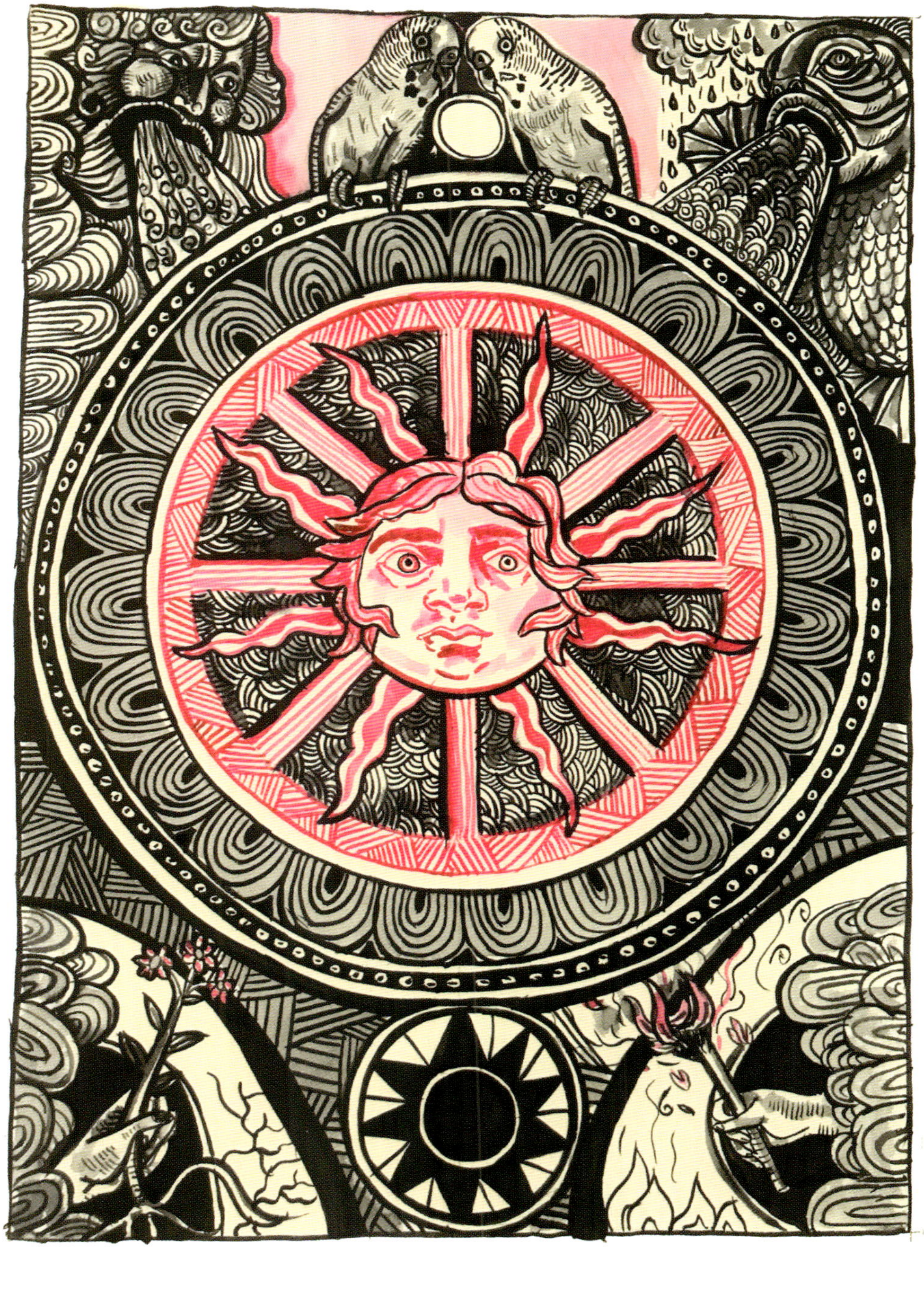

Sam Mitchell, *Blackhole sun*, 2021, watercolour, 210 × 290 mm

rogue planets

Here's some more weird information. Also known as 'free-floating planets', rogue planets are like unshackled vagabonds that have been ejected from their solar systems somehow (often as a result of their sun doing the big *kablooey!* thing and setting their surviving planets free from its gravitational pull). Apparently there are thousands of these rejects roaming around out there. The MOA telescope at the Mount John Observatory in the South Island was involved in the very first discovery of these lonely orphans.

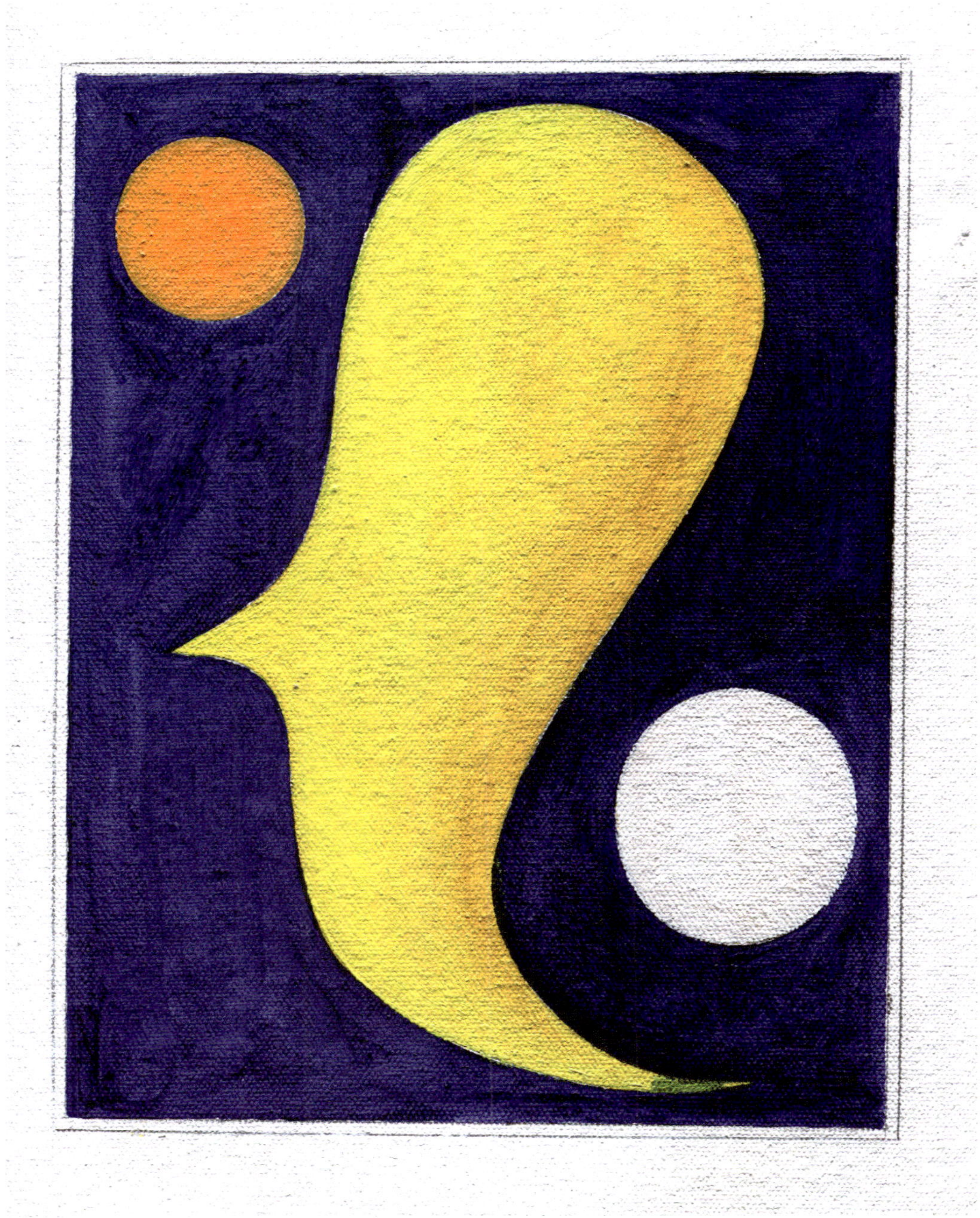

Mark Braunias, *Sun King*, 2021, acrylic on paper, 290 × 220 mm

that's enough science fact — now for some science fiction

Some people think that spending money exploring space and trying to put earthlings on the moon and Mars is a colossal waste of hip-replacement funding, but you've probably figured out by now that we can't just hang around here twiddling our thumbs. We know that we have to get out of here eventually, and we've got to start somewhere.

And there's an awful lot to be getting on with. Exploring the universe for suitable alternative accommodation is going to be a mind-bending exercise. In fact, we're going to have to change ourselves first if we're going have any hope of setting sail into what they call 'the vast reaches of outer space'.

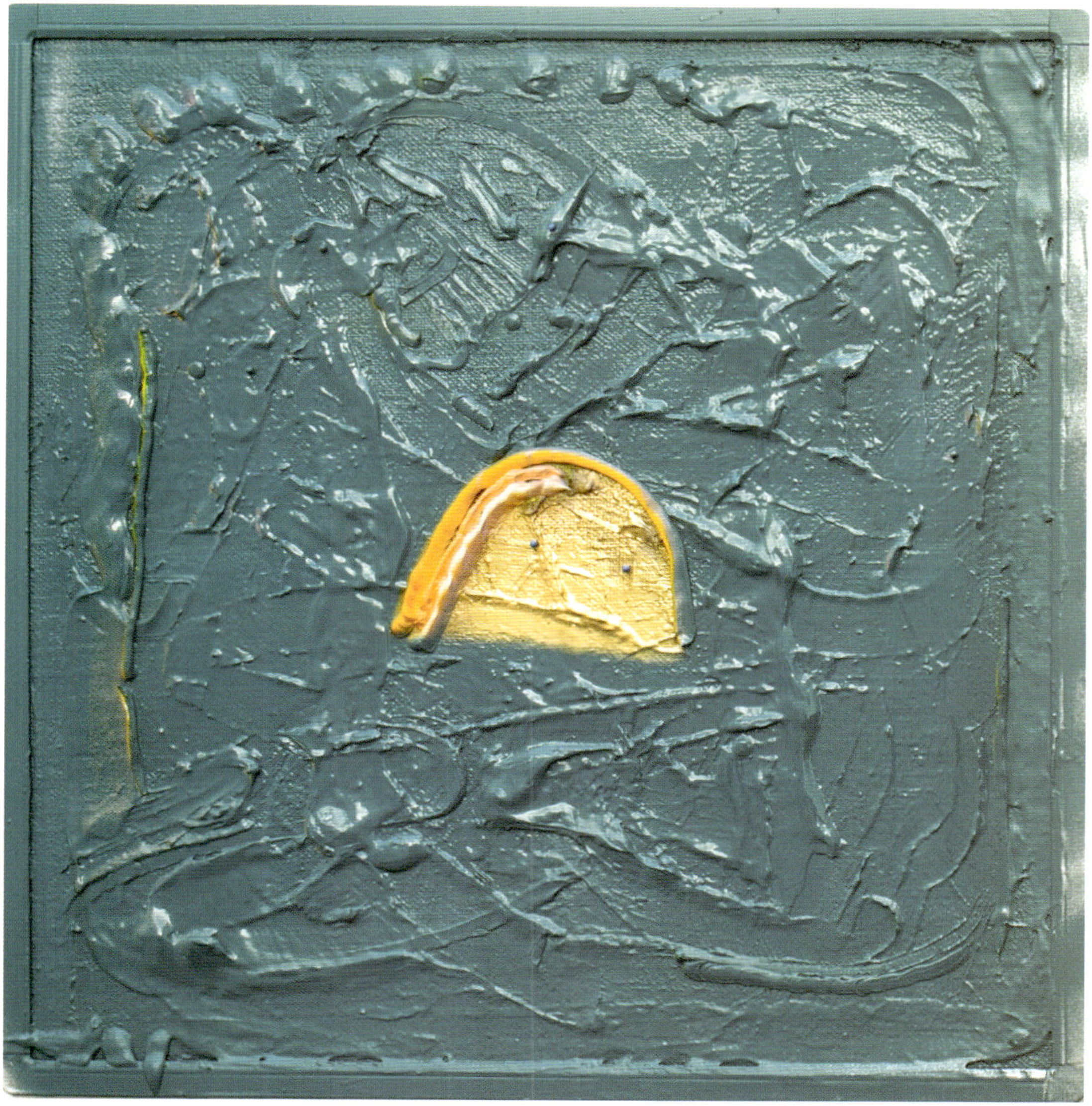

Nicholas Pound, *You have earnt your badge*, 2019, oil and spray paint on noticeboard, 700 × 700 mm

intelligence, but not quite

Big changes will have to be made if we are to face outer space and all its dangers. All the information that goes into making up our selves will have to be transferred over to something tougher than these carbon-based life-forms we currently inhabit. Sounds a bit scary — becoming robots and all that — but it's not really. There is no such thing as 'artificial' intelligence; there is only intelligence: information. AI as it exists now should be called 'Almost Intelligence' because that's what it is — intelligence, but not quite. Yet.

At some point, intelligence will cease to be artificial and become, well, it'll become us! Astronauts won't climb into their spaceships; they'll be downloaded into them. Imagine that: no freeze-dried dinners, no frozen astronauts lined up in pods on the quarterdeck.

New propulsion systems will need to be invented, too. These great drums of rocket fuel we use now aren't going to do it — there's just too much bulk.

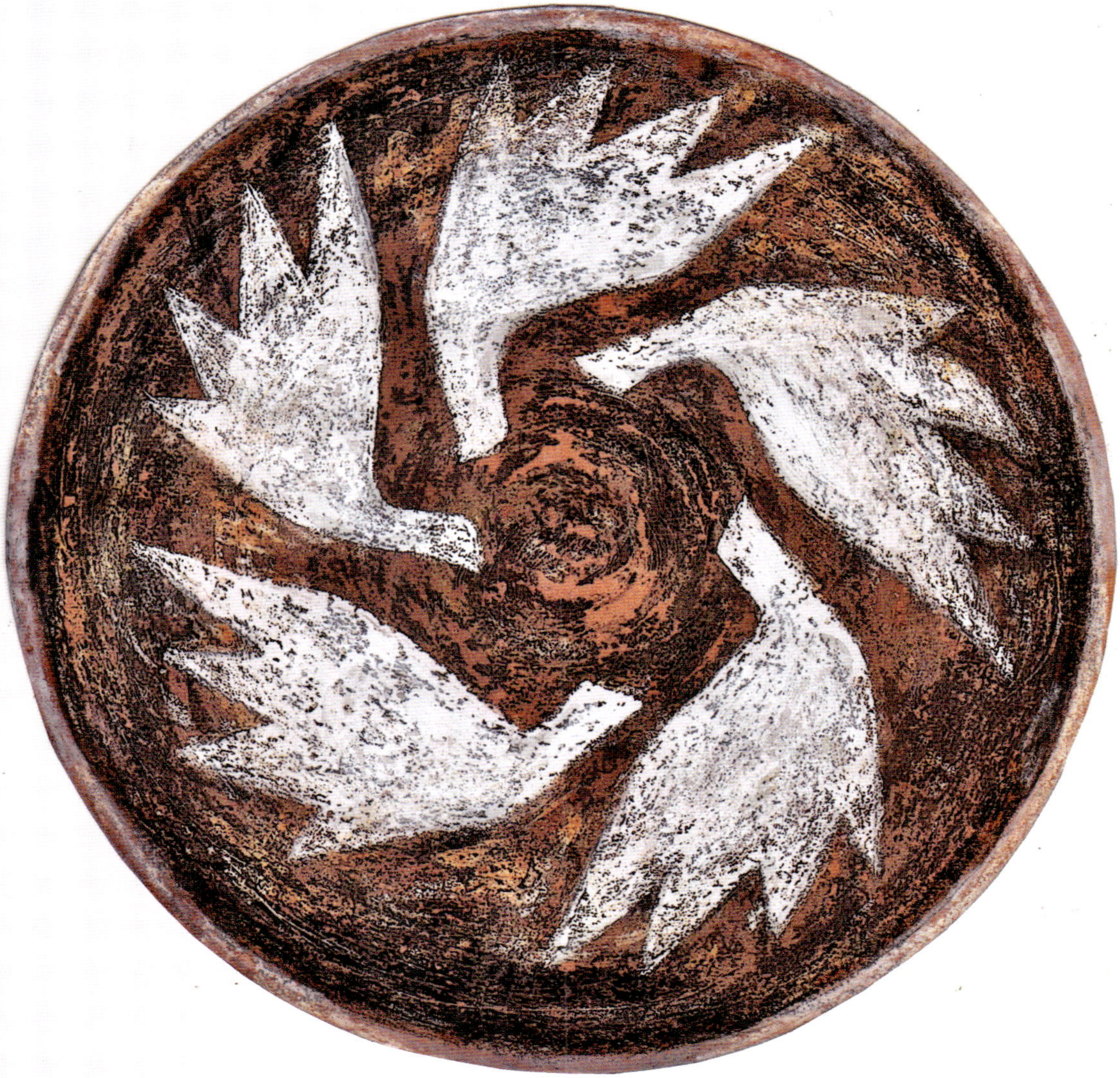

Hamish Keith and Barry Brickell, *Swan Sun*, 1961, ceramic

moving on out

I have no idea what it will be that shoots us off into the wide black yonder, but no doubt it'll be something we can harvest along the way. It might be the odd hydrogen atom hitching a ride, or all that dark matter lying about waiting to get hoovered up. Travelling in space could be a bit like those great wagon trains that opened up the American West, with the oxen chomping along on all that free grass.

And don't forget that the spaceship itself only has to be the size of a pinhead, constructed with nano-technology. It would be fired by quantum computers, guided by 'us' logged in there as pure information. If this all sounds a bit fanciful, you can be sure that it'll probably be even weirder than that. What it *won't* be is Captain Kirk sitting there on his plastic throne. I can pretty much guarantee you that.

And when we get to wherever we're going, we simply ('simply' — ha!) replicate everything we need to colonise the new place, including ourselves.

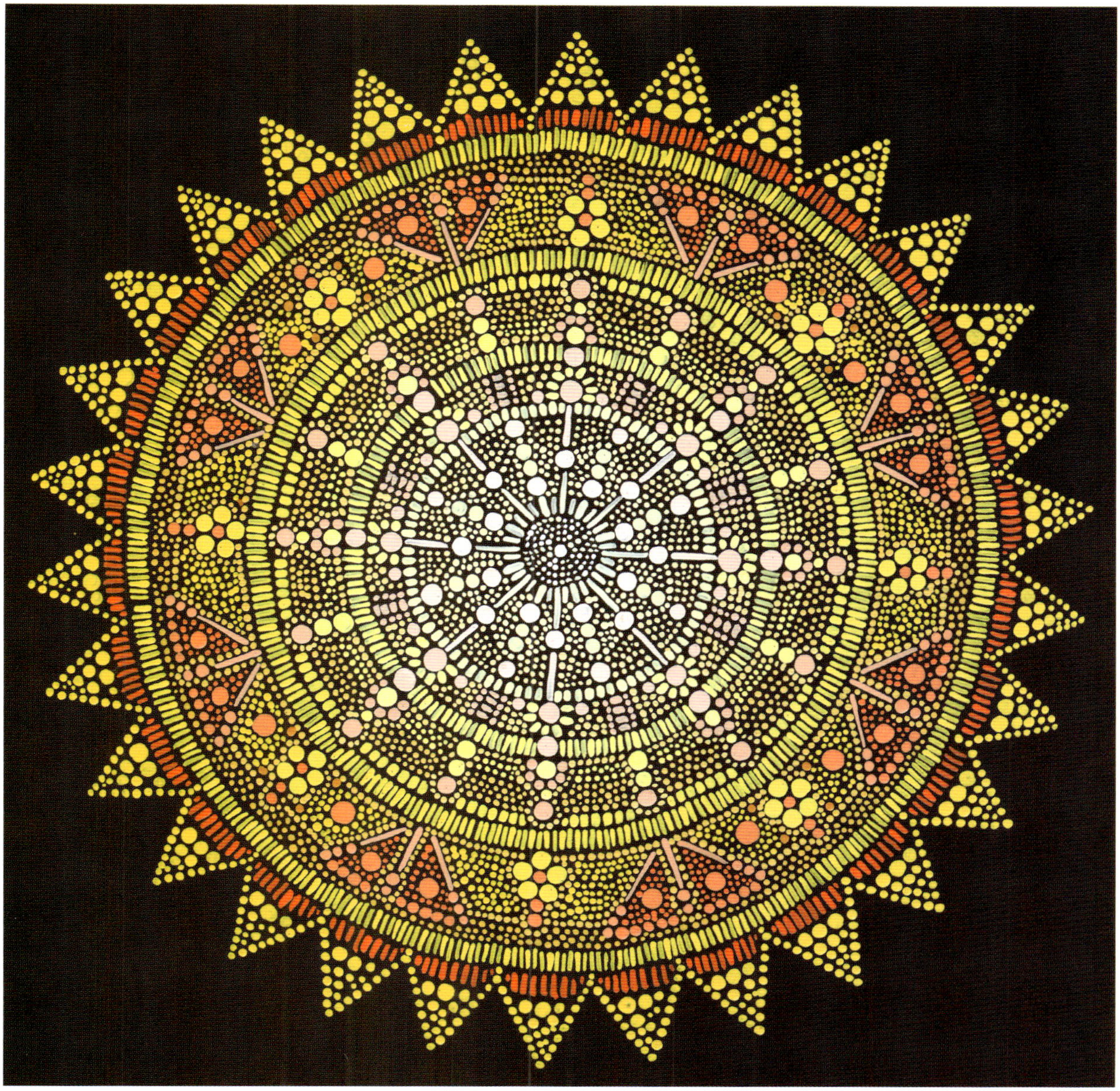

Jo Blogg, *Everybody Loves the Sunshine*, 2021, acrylic on acetate, 300 × 300 mm

we are the physical manifestation of information

The moment the microscopic universe appeared, it contained all the information needed to make us. It just took billions of years and a few exploding stars to get there. And here we are to prove it. Our appearance here on this planet is like the universe slowly waking up, opening one eye like a hibernating bear, having a bit of a look around and saying, 'Where am I? What's this all about?'

We're information on legs, basically. Self-replicating computers passing information down the great river of life, through our genes. Our genes are the information. Everything is a computer inasmuch as it is composed of its own information, its own recipe.

If everything — *everything* — is information, like DNA codes and so on, how does it work? Well, it's as if everything has a recipe that explains it, a software program it carries within it, in the same way that an acorn contains all the information needed to make an oak tree, or a fertilised egg has all the information within it to make a chicken — or you.

Even that rock I was talking about earlier is a computer, running a very simple program that says 'rock'. A rock is the very best thing there is at being a rock!

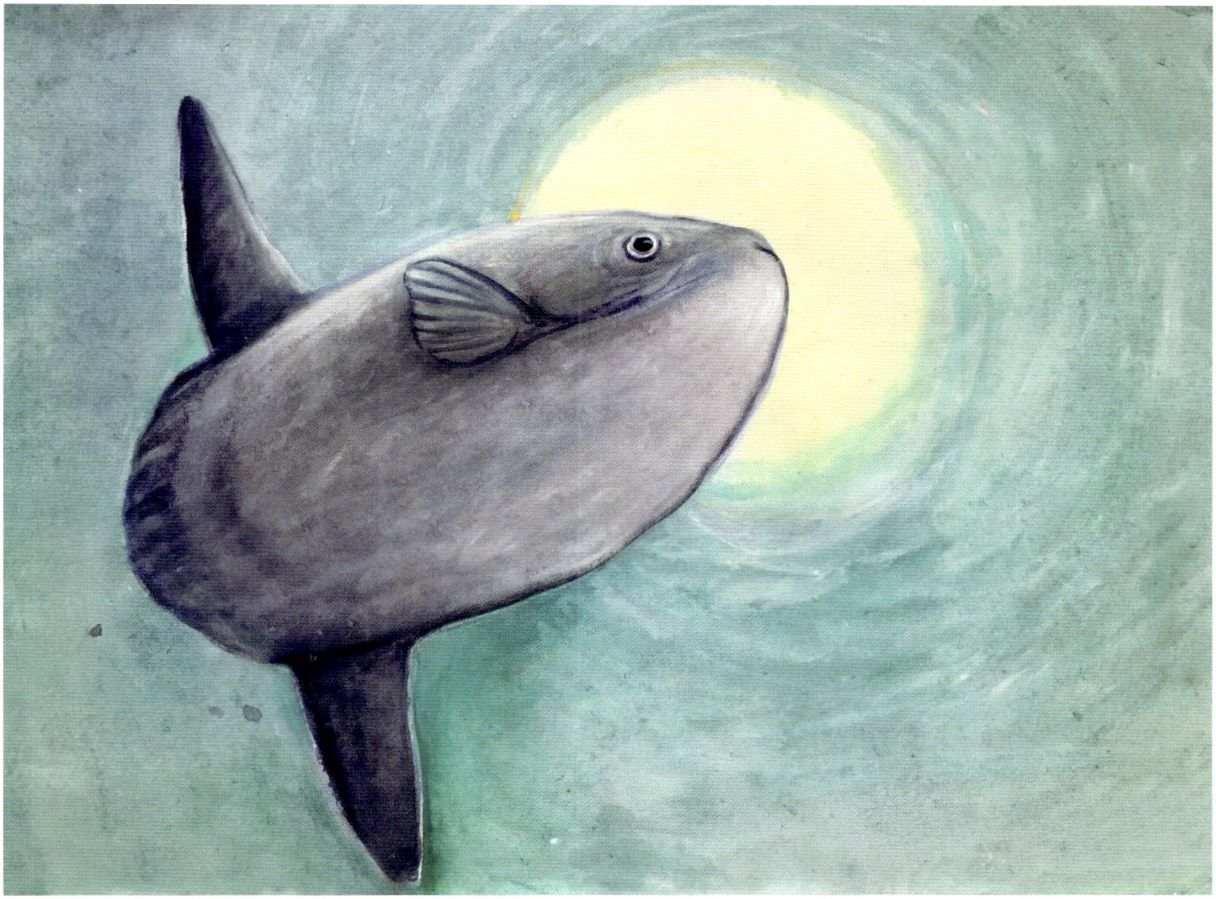

Howie Cooke, *The Moonfish and the Sun (Escape From Shimonoseki)*, 2002, watercolour on paper

the universal consciousness

Some people believe that a universal conscious mind started here on Earth. Other people say that that's all very interesting, but why would it happen here? On Earth, of all places? (They also say, 'That's the silliest thing I've ever heard of.')

I say that it had to happen somewhere, and wherever it did happen the people would say, 'Why here? It's too much of a coincidence!' That's the funny thing about coincidences. Looking back at them in the rear-vision mirror, they always look like fate.

Anyway, the idea that the universal conscious mind should have seeded *here* makes perfect sense to me. The timing, and all the information, points to it. Otherwise, what's all that information for? The universe looks very determined to explain itself. Look how far we've come already, constantly searching for meaning — all those theories, all those experiments. One thing leading to another . . . from cave-art to cellphones. Don't tell me there's no pattern or purpose there.

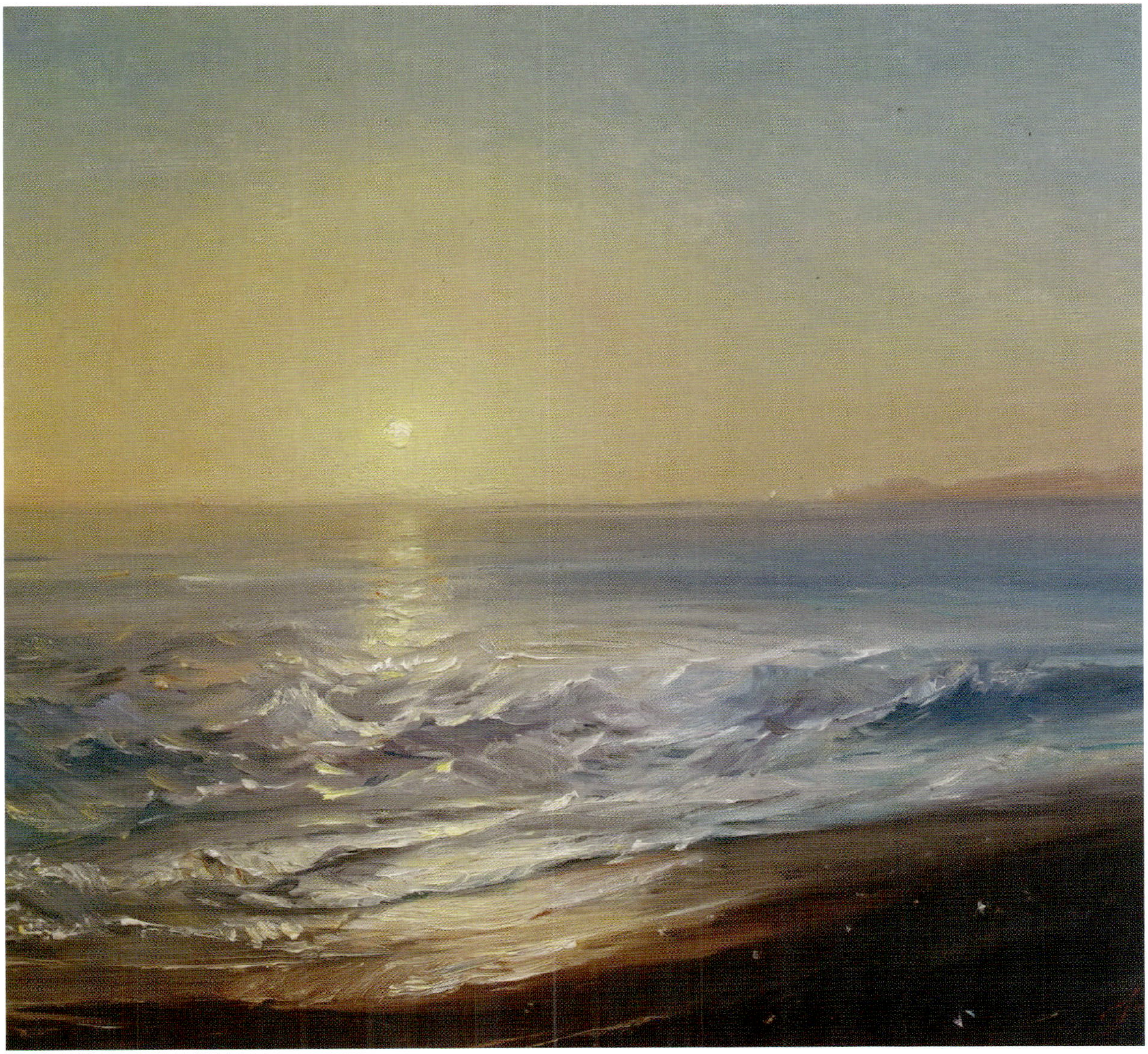

Freeman White, *Sunrise on Napier Beach* (detail), 2015, oil on panel, 200 × 250 mm

the information flow

You can see I'm just starting to make stuff up now, but you can be sure that we'll think of *something* to keep the river of information flowing. It's not going to end with just this earthly episode.

Our endless search for the ultimate theory of everything seems to be something to do with being human. So, are we the system that the universe is using to become aware of itself? When we are 'one with the universe', when we've measured and counted every last atom, we will have colonised space — and as a result we will basically *be* the universe.

Sounds bonkers, I know, but we are talking extraordinary timeframes here. The universe has existed for 14 billion years and looks like lasting for another 100 trillion more, so there's plenty of time for anything that could possibly happen to happen.

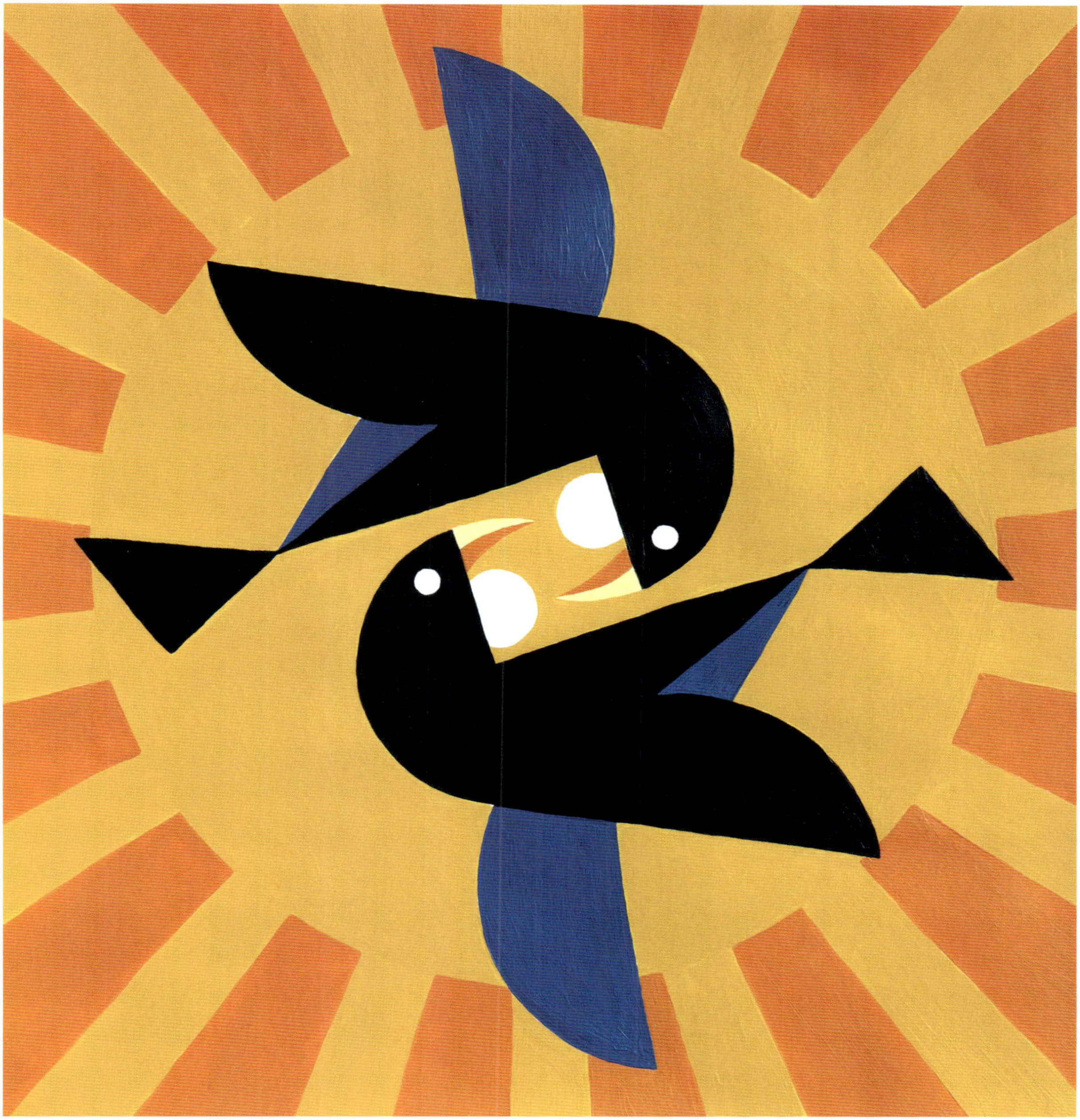

Greg Straight, *Around the Sun*, 2021, acrylic on paper, 410 × 385 mm

the meaning of life is us

Has all this happened anywhere else in this vast universe? Possibly. Who knows? There might be a Plan B. *We* might be Plan B!

It's a waste of time thinking about it, really. If there's life out there somewhere we'll find it one day, but in the meantime let's get on with it. Let's keep reaching for those stars!

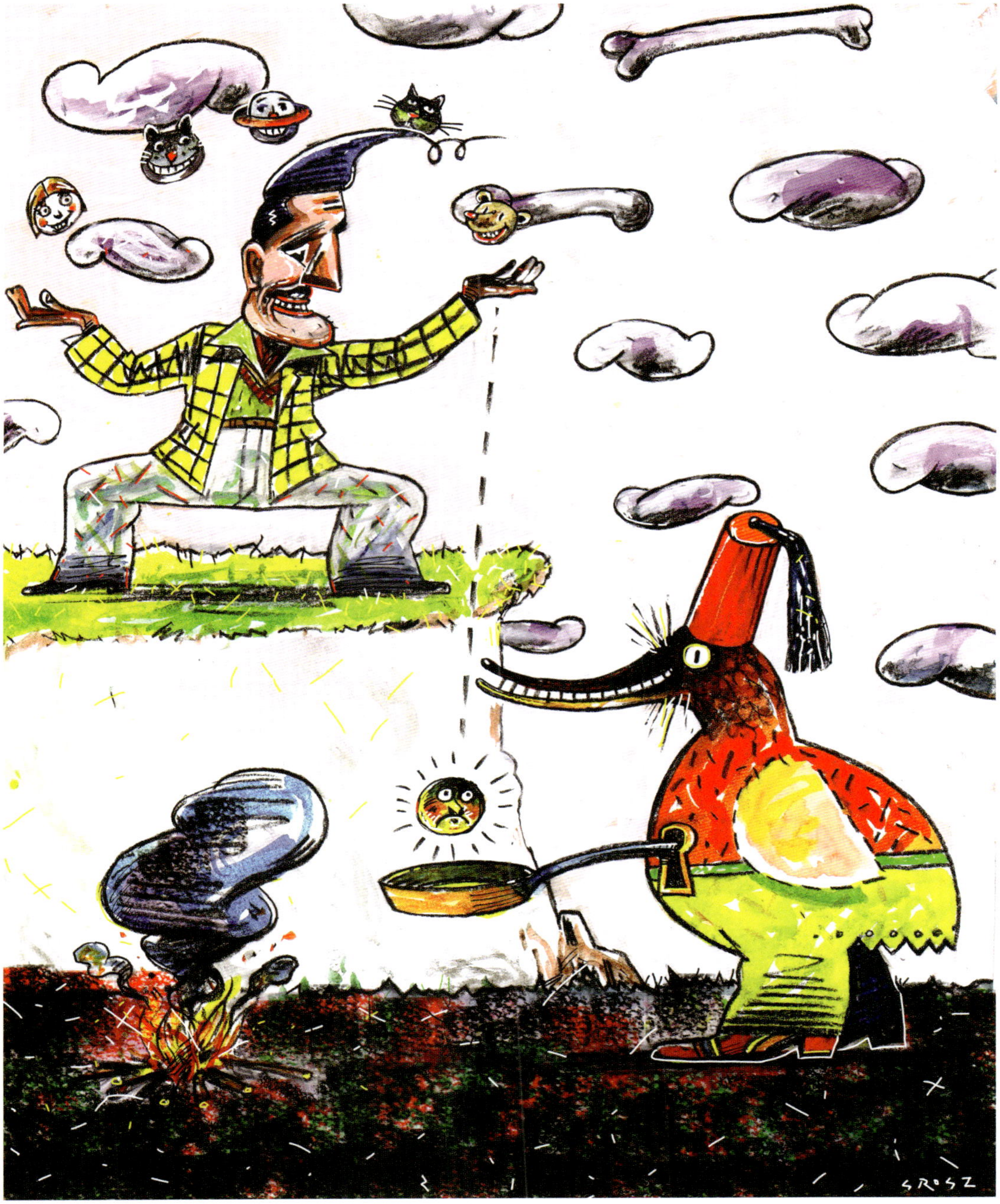

Chris Grosz, *from the climate change frying pan into the climate change denialisms fire*, 2021, charcoal, gouache and collage, 300 × 360 mm

a short glossary

Big Bang theory The idea of the universe expanding out propulsively from a singular moment of nothing was first put forward by the Belgian priest and theoretical cosmologist Georges Lamaître back in the 1920s. The English astronomer Fred Hoyle introduced the term as a sort of scientific joke way back in 1949, but it caught on and soon became the universally accepted way to describe the expansion of the universe from an initial state of high density and temperature. Of course, there wasn't a bang, and, lasting only for a nanosecond, it wasn't very big either, but something dramatic certainly happened. Everything we know and everything we don't know came from nothing 13.8 billion years ago and, despite what people might try to tell you, we simply haven't got the science to explain what that 'nothing' was.

centrifugal force This is the apparent force that an object like you feels when you go round a corner in a car. You feel flung out from the centre of the circle you're turning. It's the sort of trade-off between gravity and centrifugal force that keeps planets in their orbit.

cosmology This is the 'Big One'. The whole ball of wax. The study of physics is the examination of matter and energy. Astronomy is all about the galaxies, the stars and all their possible planets. Cosmology is about where it all might have come from and where it all might be going.

density You can go totally scientific on this one, but it's basically about how compacted something can get, and its comparative weight. Here's an example: if you bought two orders of candy floss at the A&P show and you squished one down to a nasty little plug of sugary gunk it would still weigh the same as the huge pile of sugary fluff in your other hand. This is what happens when a sun blows up and collapses back to dwarf status — it still weighs about the same as it did before it compacted! We all know that great fact about a teaspoon of a white dwarf star weighing 15 tons, don't we?

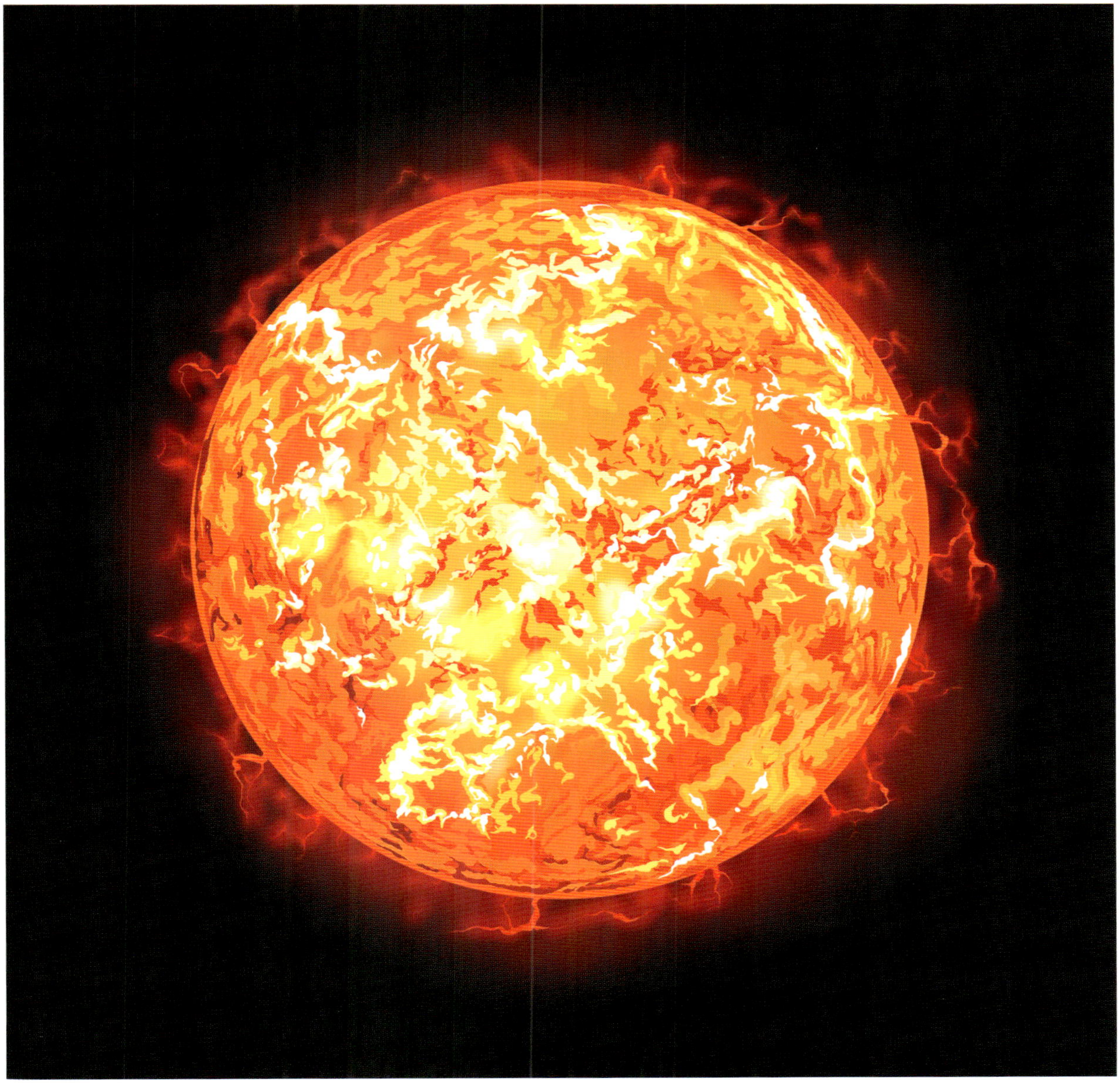

Otis Frizzell, *Digital Illustration*, 2021

galaxy We all know what this one is, don't we? 'In a galaxy far, far away' and all that? It's like an island in a huge, dark ocean, separated from other islands by distances so vast that they're almost impossible to comprehend. All the talk of space that we usually hear — solar systems with possible habitable planets, spiral arms, black holes, white dwarfs — are all in our own backyard, the Milky Way. Detailed research on other galaxies is, for now, literally beyond our reach.

light year A light year is the distance that light travels in a year. Which sounds pretty straightforward . . . except for the speed in which it does it, which is — wait for it — 299,702 kilometres a second. A *second!* That adds up (so I'm told) to 9.46 trillion kilometres a year. I'm going off for a wee lie-down.

oblique angle Imagine opening a door past the point where it's sticking straight out from the hinges and it's on its way to being flat against the wall: that's an oblique angle. If you move the door as if you were shutting it, the angle becomes acute as it closes in on the latch.

photovoltaic I can't really explain the actual photovoltaic effect but I do know that it works in those big black glossy solar panels you see everywhere, where the energy from the sun is converted into a flow of electrons, which is, conveniently, what electricity is! This could catch on!

plasma This is a good one. Plasma is the fourth state of matter. Simply put (ha ha), it's an ionised gas with the remarkable ability to free up electrons and allow them to leave their atomic bondage and make new friends with ions. I'm not sure what ions are either, but if there is indeed only one electron in the entire universe, then it needs all the friends it can get.

space–time A mathematical model which fuses the three dimensions of space with the fourth dimension of time. Space is the 'stuff' and time is, well, time. One cannot exist without the other, hence the conflation of the two terms into one. It is basically all one thing made up of two things — right? Just as you can't have a table with only the legs, or only the top — it's not a table until the two systems are brought together — time exists in things, and things exist in time.

universal consciousness An idea first put forward by 'ancient eastern mystics' — and The Beatles — which posits that if we just concentrated hard enough our individuality would sort of disappear in a purple haze of spiritual commonality. Whether this is meant to happen at a sub-atomic level that hasn't yet been identified is hard to qualify, but when the information we all carry within us is indexed directly and seamlessly to the information that defines the very structure of matter, the result won't be universal consciousness so much as a conscious universe.

vector This is how we measure one point in space relative to another — like those pin boards you see in TV thrillers where the detectives have used thread to link points of interest, but in space. The vector point is arrived at by triangulation, but if you'd said strangulation I would have still believed you. It really is a great 'spacey' word and gets thrown around pretty much willy-nilly.

warp and weft These are the terms used to describe the process of weaving yarn into fabric. We talk so much about the 'fabric of space and time' that it just follows that we should adopt warp and weft to describe this difficult concept. It's interesting that we even use the term 'time warp' to describe an imaginary rip in this fabulous fabric. Unfortunately, we have yet to find any evidence of time warps and wormholes.

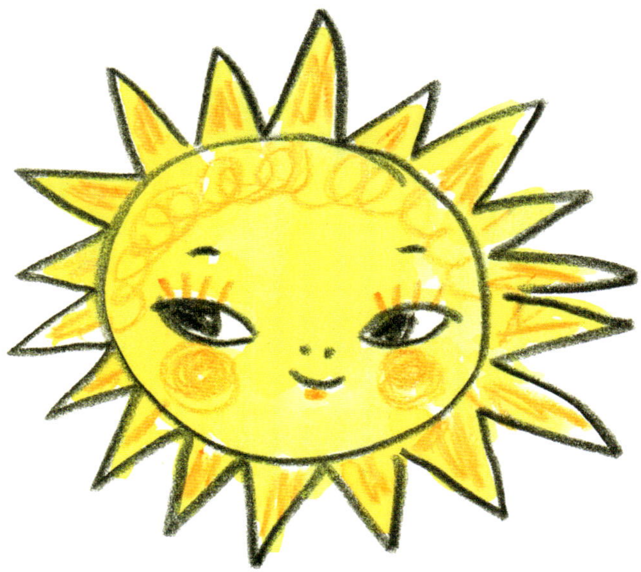

Misery, *Sunny*, 2021

about the authors

Dick Frizzell MNZM is one of New Zealand's best-known painters. He studied at the Ilam School of Fine Arts at the University of Canterbury from 1960 to 1963 and then had a long career in advertising. Alongside his career as a painter, Frizzell is also the highly sought-after designer of a range of products from toys to wine. He is the author of *Dick Frizzell: The Painter* (Random House, 2009); *It's All About the Image* (Random House, 2011); and *Me According to the History of Art* (Massey University Press, 2020). Dick exhibits regularly and often works in collaborations with writers and other artists. He lives in Auckland with his wife, Jude.

Samantha Lord is a passionate science communicator who rediscovered her love of astronomy when she arrived in New Zealand and started living and working in the Aoraki Mackenzie International Dark Sky Reserve. She believes that engaging in astronomy provides the opportunity to probe the vastness of the cosmos whilst also giving a sense of context that enables us to reflect on life on our own planet Earth.

Having completed a degree in Chemistry at the University of Bristol, Lord worked for nearly four years as an astronomy guide at Lake Tekapo. A keen writer, she finished a Postgraduate Certificate in Science Communication through the University of Otago in 2020. She has recently returned to England to work as an engagement officer at Jodrell Bank Observatory, helping to share the science undertaken by the historic Lovell Telescope — the third-largest steerable telescope in the world.

and the artists

Jo Blogg is a Hawke's Bay-based artist who works in multiples and patterns to explore issues including feminism, consumerism and environmental degradation. She holds a BA with Honours from the London Guild Hall University, and her work has been shown both nationally and internationally.

Richard Boyd-Dunlop lives in Hawke's Bay, where he opened Boyd-Dunlop Gallery in 2014, first as a space to exhibit his own work and now as a hub for contemporary art in the region. Known for his bold use of colour and large-scale works, Boyd-Dunlop is a self-taught artist with a career spanning over 25 years. His work is held in the Auckland Regional Council collection, the Manukau City Council collection and many private collections both in New Zealand and around the world.

Mark Braunias graduated from Ilam School of Fine Arts in 1988. The inaugural winner of the James Wallace Art Award in 1992, he has received various scholarships and residencies, including a Wallace/Fulbright scholarship to San Francisco in 2011, a William Hodges Fellowship, Invercargill (2005), and residencies at the Dunedin Public Art Gallery (2002) and Tylee Cottage, Whanganui (2007). Known for his sinuous and bulging biomorphic paintings and drawings, the artist has also collaborated with printmakers, animators, quilt makers and glass artists. Braunias lives and works in Kawhia, Waikato.

Howie Cooke is a visual artist who has exhibited extensively over the last 45 years. Originally involved in the Auckland art scene of the 1970s and 1980s, Cooke now resides in Byron Bay, Australia, where he is immersed in art, music and an ongoing celebration of the Oceanis. He has painted large whale murals throughout the world, often in conjunction with international whaling commissions, Sea Shepherd and his own group Surfers for Cetaceans, a non-profit organisation created to encourage action against whaling and dolphin kills.

Judy Darragh ONZM lives and works in Auckland and has played a significant role in the development of Artspace (now Artspace Aotearoa), and the independent artist-run spaces Teststrip and Cuckoo. Her witty and provocative work, typified by brightly coloured sculptural assemblages of found objects, recycled items, industrial materials, collage, photography, video and poster art, displays a fondness for the everyday. Darragh has taught art at secondary and tertiary levels for over 40 years and has mentored young artists throughout Aotearoa.

Susan Edge lives in Waipu, north of Auckland. She paints quirky pictures in acrylic and dabbles in collage. She exhibits in group shows, has held several solo exhibitions and enjoys commissions. Her work is held in collections throughout New Zealand and abroad.

Fane Flaws is a songwriter, musician and artist. He graduated from Wellington Polytechnic with a Diploma in Graphic Design in 1971, and then worked primarily in the music industry with bands including Blerta and Spats, which later morphed into The Crocodiles, before returning to painting with his first solo exhibition in 1984. He has been inducted to both the Massey University Design Hall of Fame and the New Zealand Film Archive Music Video Wall of Fame.

Otis Frizzell is Dick Frizzell's son, and is also an artist. Go figure.

Max Gimblett ONZM is a painter, calligrapher and Rinzai Zen monk. His work, a harmonious post-war synthesis of American and Japanese art, brings together abstract expressionism, modernism, spiritual abstraction and Asian calligraphy. Gimblett is represented in the collections of the Guggenheim, the Museum of Modern Art, the Whitney Museum of Art, the National Gallery of Art (Washington, DC), the San Francisco Museum of Modern Art, Museum of New Zealand Te Papa Tongarewa, the Auckland Art Gallery Toi o Tāmaki, the Queensland Art Gallery, Gallery of Modern Art, Brisbane, and the Art Gallery of New South Wales, Sydney. He lives and works in New York and regularly visits New Zealand.

Chris Grosz is a poster designer, political and editorial illustrator, writer and musician. Grosz and Dick Frizzell were going to start an illustrator duo in 1969 called Ink Inc. It didn't happen, but Grosz went on to teach graphic design and pursue image creation in various forms, including illustration, film and animation work. He illustrated a long-running series of columns by Shane Maloney in Australian magazine *The Monthly*, which became an animated series for the ABC. Having lived in Melbourne for 30 years, Grosz is back in New Zealand illustrating a regular Audioculture Iwi Waiata column, 'Rock & Roll Rendezvous', with Chris Bourke.

Hamish Keith CNZM OBE is an artist, art commentator, art historian, art curator, television presenter and columnist. Co-author, with Gordon Brown, of the landmark publication *An Introduction to New Zealand Painting 1827–1967*, Keith is also well known for his television series *The Big Picture*, which covered New Zealand art from 1642 through to the contemporary. Keith worked at the Auckland Art Gallery from 1958 to 1970, and has served on various arts sector boards, including the Arts Council, the Council of the National Art Gallery, the Museum of New Zealand and the board of the Auckland Art Gallery.

Phil King was born in Bristol in 1965 and spent his early childhood in Toulouse, France. He completed a BA in Fine Art Painting at Bath Academy of Art, and then an MA in Fine Art at

Goldsmiths College, London. King's career as an artist has spanned over 20 years and has included solo and group exhibitions in the United States and the United Kingdom. As well as painting, King serves as the co-editor of *Turps* magazine, and frequently writes critical essays for exhibition catalogues.

Peter Kingston received a Bachelor of Arts from the University of New South Wales in 1965. A celebrated landscape painter, his survey exhibition *Habourlights* was toured by the Manly Regional Gallery in 2004. Kingston has also been included in numerous group exhibitions, including the Wynne Prize (between 1995 and 2003), and the Sulman Prize (between 1998 and 2001). In 2019, his work featured in *Bohemian Harbour: Artists of Lavender Bay* at the Museum of Sydney.

Virginia Leonard graduated with a Master of Fine Arts in painting from Whitecliffe College of Arts and Design, Auckland, in 2001. In 2013 she began working with ceramics. Her first major ceramic project was for craft and design gallery Objectspace and culminated in the 2015 solo-exhibition *The Effects of Crack*. Leonard's work is held in private and public collections in New Zealand, Australia, Miami, Italy and Switzerland.

Karl Maughan studied at Elam School of Fine Arts, graduating with a Master of Fine Arts in 1987. Best known for his lush, hyper-real garden-scape paintings, his work has been exhibited widely both nationally and internationally, with notable shows including *Saatchi Collection Catalogue Show* (Saatchi Gallery, London, 1998), *Stop Making Sense* (City Gallery, 1995) and the solo exhibition *Walking in Light* (Vertigo Gallery, London, 2003). Maughan returned to New Zealand from London in 2005 and currently lives and works in Wellington.

Tanja Jade McMillan (Misery) is a pop artist whose work ranges from paintings, sculpture and wall art to animation, fashion, toys and the children's decor brand Misery Guts. Her works reference her Tahitian/Chinese heritage, evident through her subject matter with their flattened depth of field, her mode of storytelling and her ink work. She lives and works in Grey Lynn, Auckland.

Sam Mitchell was born in Colorado Springs in 1971 and currently lives in Auckland. Mitchell studied at Elam School of Fine Arts, gaining a Bachelor of Fine Arts in 1997, and is known for hi-gloss, candy-coloured perspex portraiture. Mitchell won the Paramount Prize in the Wallace Art Awards in 2010 and has been awarded several national and international residencies, including the International Studio & Curatorial Program (ISCP), New York (2010); The Lockup, Sydney (2010); the William Hodges Fellowship, Invercargill (2014); and the Tylee Cottage artist residency, Whanganui (2015). The artist's work is held in many private and public collections both in New Zealand and internationally.

Gregory O'Brien is a poet, essayist, curator, painter and printmaker. His most recent book is the 'oceanic sketchbook' *Always Song in the Water* (Auckland University Press, 2019). An exhibition of his recent paintings, *The Wading Birds of Drybread*, toured the South Island in 2019–20. He lives in Wellington and has exhibited at Bowen Galleries, Wellington, for 30 years.

Chris O'Doherty, aka Reg Mombassa, was born in Auckland in 1951 and lives in Sydney, Australia. He has exhibited paintings, drawings and prints at Watters Gallery since 1975, and now exhibits with Rogue Pop-up Gallery. Mombassa is a guitar player, singer and songwriter, and has been a member of bands Mental as Anything (1976–2000), Dog Trumpet (since 1991) and The Pinks (since 2012). He designed t-shirts and posters for Mambo Graphics from 1986 to 2006, and has designed posters, book and CD covers, and logos for a variety of clients.

Ani O'Neill, of Cook Island and New Zealand heritage, graduated from Elam School of Fine Arts in 1994 and has exhibited throughout New Zealand, Australia and Asia. Working in crochet, O'Neill brings together the cultural traditions of the Cook Islands, Rastafarianism and American Pop Art to address issues such as post-colonialism, consumerism, identity politics and institutional hierarchies within the arts sector. She lives in Rarotonga and returns regularly to New Zealand.

Reuben Paterson (Ngāti Rangitihi, Ngāi Tūhoe, Tūhourangi) graduated from Elam School of Fine Arts in 1997 and in the same year received the Möet et Chandon Fellowship to Avize, France. Known for his glittering artworks that frequently incorporate kōwhaiwhai patterns, geometric tessellations and floral motifs, Paterson explores such themes as memory, whakapapa, nature, life, death and rebirth. Paterson's work is held in major collections in New Zealand and abroad, and has been included in major exhibitions, including the 8th Festival of Pacific Arts Biennale d'Art Contemporian, Nouméa, New Caledonia; the International Biennale of Contemporary Art at the National Gallery in Prague; the Asia Pacific Triennial in Queensland; and the 17th Biennale of Sydney, Australia.

Martin Poppelwell studied at Elam School of Fine Arts in the early 1990s and later under potter Ross Mitchell-Anyon, gaining a Diploma of Ceramic Design and Production at Wanganui Polytechnic. A renowned potter, Poppelwell has also worked in printmaking, design, film, fashion and textiles. His solo exhibitions include, most recently, *meaninglessnessless* (Melanie Roger Gallery, 2020) and *the edge of the alphabet* (Jonathan Smart Gallery, 2019). His work is held in collections in New Zealand and abroad, including the Museum of New Zealand Te Papa Tongarewa, the Lawrence B. Benenson Collection, New York, US, and the Reverend Ian Brown Collection, Melbourne, Australia. He lives in Hawke's Bay.

Nicholas Pound graduated with a First Class Honours degree from Elam School of Fine Arts in 2015. Pound studied American poetry prior to painting, and this interest is retained in his work, which often employs language-poet strategies, and explores interpretation, idiosyncrasy and malapropism.

Patrick Pound lives and works in Melbourne, Australia. Pound's work treats collection as a medium, ordering photographs and objects around an idea or proposition. In 2020 one of his large-scale installations of intersecting photographic images featured in the Biennial for Contemporary Photography at Kunsthalle Mannheim in Germany. The same year, Pound was commissioned to respond to the Art Gallery of New South Wales's collection in *Shadow Catchers*, curated by Isobel Parker Philip. The gallery also holds the artist's assemblage of 120 photographs of suns, moons and numerous other spheres, titled *The Image Pool* (2016). A major career survey, *Patrick Pound: The Great Exhibition*, was held at the National Gallery of Victoria in 2017.

Thomas Newman Pound trained as a theatre/set designer and worked as a contract display artist at Auckland War Memorial Museum Tāmaki Paenga Hira before committing to full-time art practice. He often works with found mixed media. The work *9027* is from his solo show *The Instance*, held at Government House, Auckland, in 2018. Represented by Emde Gallery, Mainz, Germany, Pound lives and works in Auckland.

John Pule ONZM was born in Niue and has lived in Auckland since 1964. His work explores the mythologies, histories and colonisation of the Pacific, and draws on themes of migration and the impact of Christianity on indigenous cultures. His art has been exhibited widely, including in *Paradise Now?* (Asia Society and Museum, New York, 2004) and the 2006 Asia-Pacific Triennial (Queensland Art Gallery, Brisbane, Australia). In 2011 the Auckland Art Gallery Toi o Tāmaki held a major survey show, *John Pule: Hauaga (Arrivals)*.

John Reynolds graduated from the Elam School of Fine Arts in 1978. Best known as a painter, who draws on abstract expressionism or text, Reynolds has also produced public artwork, sculpture, installation and temporary or site-specific works. His solo exhibitions include *The Violet Hour* (Auckland Art Gallery Toi o Tāmaki, 2019), *Manifesto* (Tauranga Art Gallery Toi Tauranga, 2015) and *Epistomadologies* (Auckland Art Gallery Toi o Tāmaki, 2015). In 2006 his large installation work *Cloud* featured at the Sydney Biennale at the National Art Gallery of New South Wales. He has received numerous awards, including the Montana Lindauer Award (1988), and a Visual Arts Fellowship from the Arts Council of New Zealand (1993).

Simon Shepheard graduated from Elam School of Fine Arts in 1983 and is best known for his work with found materials, including high relief paintings, animation, kinetic and interactive pieces. He has had work featured in major group exhibitions, including *Transformers* (Auckland Art Gallery, 1996; City Gallery, Wellington, 1996–97), *White on White* (Christchurch Art Gallery Te Puna o Waiwhetū, 2008–09) and the Wallace Art Awards touring exhibition (nationwide, most recently 2017–18), and his art is held in various local and international collections. Shepheard tutored and lectured at Elam from 2000 to 2005 and currently divides his time between Auckland and the Tutukaka coast.

Greg Straight is a graduate of Elam School of Fine Arts, where he was inspired by the work of his tutor Dick Frizzell. He is well known for his commercial illustration work for high-profile brands, which includes murals, packaging designs, billboards and magazine covers. His successful career in graphic design led him to a six-year stint in London before returning to Aotearoa, where since 2010 he has produced a range of art prints playing on Kiwi emblems such as summer roadies, riding bikes or hitting the waves. In these, style and hues are tweaked and reworked and results are unexpected and intelligent.

Grahame Sydney ONZM is a painter, printmaker, photographer and writer, well known for his depictions of Central Otago. His work is held in public institutions including the Museum of New Zealand Te Papa Tongarewa and private collections in New Zealand and throughout the world. He is the author of several books, including *Grahame Sydney Paintings: 1974 – 2014* (2014); *Timeless Land*, with Brian Turner and Owen Marshall (1995); *White Silence: Grahame Sydney's Antarctica* (2008); *Grahame Sydney's Central Otago* (2021); and, most recently, with Brian Turner and Owen Marshall, *Landmarks* (2020). He lives and works near St Bathans, in Central Otago.

Darryl Thomson
Ko Puketapu te maunga
Ko Ngaruroro te awa
Ko Omahu te Marae
Ko Ngāi Te Ūpokoiri/Ngāti Hinemanu ngā hapū
Ko Ngāti Kahungunu te iwi

As a descendant of Maui, who captured and slowed Tamanui-te-rā down to create longer days, Darryl Thomson is constantly reminded by the big skies and long summers of the tribal area of Ngāti Kahungunu of his tipuna's feat. Tamanui-te-rā is a source of light, health and a reminder of their living relationship with their tipuna and environment. He loves painting Tamanui-te-rā because there's nothing better than bringing warmth to a place by painting him.

Henry Turner is a Christchurch-based artist. Currently working in gouache and sculpture, his oeuvre has included fruit-like forms, vibrant landscapes, strange things and apocalyptic symbols. Turner has exhibited widely in New Zealand, including, in 2020, a solo exhibition, *Fever!*, at City Art Depot, Christchurch.

Denys Watkins is a highly respected printmaker and painter with a career spanning five decades. He attended the School of Design, Wellington Polytechnic (1962–64), Central School of Art, London (1967), and the Royal College of Art, London (1968–70), and went on to teach at Elam School of Fine Arts from 1980 to 2011. He has been awarded multiple scholarships throughout his career and was the recipient of the Artist in Residence programme at the Canberra School of Art in 1985 and the Sanskriti Foundation Residency, India, 2003. Throughout his career he has drawn on a diverse range of inspirations for his abstract paintings and prints, including Hindu monkey gods, coral, clouds, botany and song lyrics. A major retrospective of his paintings from 2004–16 toured galleries throughout New Zealand in 2018–19. Watkins lives and works in Auckland.

Weston Frizzell is the collaborative identity of artists Otis Frizzell and Mike Weston. The 'hi-performance art partnership' has its roots in the street art and pop culture melting pot of Auckland's K Rd of the late 1990s. Their collaboration applies a production-team approach to producing one-off and edition painted works, an extensive range of limited-edition screenprints and digital works, and a range of mass-marketed merchandise. Their work is informed by Dada, punk, situationism, hip-hop, graffiti art and techno culture, and juxtaposes art history with contemporary media, often with a political message at the centre.

Freeman White shot to national and international acclaim after winning the Adam Portraiture Award in 2006 at the age of 26. He is recognised as one of New Zealand's leading landscape and portrait painters and has also worked on films, including *The Hobbit* and *What We Do in the Shadows*. His work is held in various public collections, including the New Zealand Portrait Gallery Te Pūkenga Whakaata and Wellington Museum. A father of two, White lives and works in Hawke's Bay.

Wayne Youle (Ngāpuhi, Ngāti Whakaeke, Pākehā) lives and works in Amberley, North Canterbury. A graduate of the Wellington Polytechnic School of Design, his work is often humorous and addresses issues of identity, race and the commodification of cultural symbols. Youle's work has been shown in national museums and public galleries throughout New Zealand and overseas, and is held in numerous collections, including the Museum of New Zealand Te Papa Tongarewa, Auckland Art Gallery Toi o Tāmaki and Christchurch Art Gallery Te Puna o Waiwhetū.

acknowledgements

I'm not sure how I ever thought I could write a book — even a simple book — about astronomy and cosmology without any help. I just think I thought I knew what I was talking about. God knows, I *felt* well-informed enough! And, you know . . . fools rush in.

I ran the first draft past my grandson Jimi Frizzell, who shares a similar interest in the infinite mysteries of time and space, and he sent the manuscript back looking like a lace curtain. I stitched the vastly improved document back together, enough to get my fabulous publishing friend Nicola Legat at Massey University Press interested. Then Jude and I took off for the South Island leg of the promo tour for my last book on the history of art.

My friend Christine Fernyhough, who knew I was writing about the stars, insisted that we stop off at Lake Tekapo and 'do' the Mount John Observatory tour — which we duly did, only to be foiled by the weather. Instead of looking through the telescope, I sat in on a lecture in the auditorium, where a very well-informed young astronomer called Samantha Lord kicked off her talk with a startling iteration of the first sentence of this book. This made me sit up and pay attention. She then proceeded to tell me things about the sun and the stars that I'd never heard before! Oops!

Perhaps some professional help would be handy right about now? So I contacted Samantha, who was only too happy to come on board as the resident expert. And what an expert she turned out to be, sorting out all my dodgy research and dizzy interpretations. Her assistance has been invaluable. I'm also hugely grateful to another associate of the observatory, Graeme Murray MNZM, who gave the project a quiet endorsement just when I needed it. With the thumbs up from Graeme, I felt ready to go.

And finally, the last person to climb into my highly scientific researches was Teresa McIntyre, copy editor, who was charged with making sense of it all. Even with Samantha's approval I was still nervous about some of my 'interpretations' . . . and so I should've been. But Teresa sorted them, and all is good. Copacetic, even.

It is with a huge sense of relief and gratitude that I thank all my artist friends for so generously getting into the spirit of the project and providing the brilliant 'illuminations' throughout the text. There was no fee: all I promised was glory and 'fulsome acknowledgement' — and this is all they get!

And, of course, heartfelt thanks to everyone at Massey University Press for being brave enough to humour me on this one. It's been a joy. As usual.

index

Day into night, night into day,
on and on, on and on.

— Turpsy Griddler, 1983

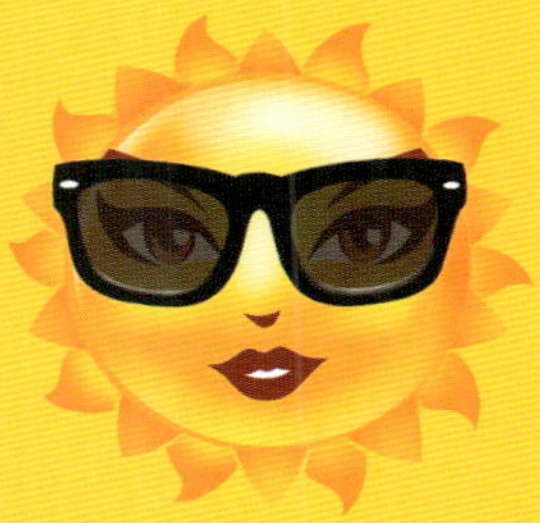

MASSEY UNIVERSITY PRESS

First published in 2021 by Massey University Press
Private Bag 102904, North Shore Mail Centre
Auckland 0745, New Zealand
www.masseypress.ac.nz

Design by Kate Barraclough
Cover artwork: Dick Frizzell, *Celestino*, 2021
Endpapers: Primary school teacher's 'excellent' stamp, c. 2000
Page 103: Otis Frizzell, *Sun emoji*, 2021

A catalogue record for this book is available from the National Library of New Zealand

Printed and bound in China by 1010 Printing Asia Limited

ISBN: 978-0-9951465-6-3